Darwin's Dogs

How Darwin's pets helped form a world-changing theory of evolution

Emma Townshend

FRANCES LINCOLN LIMITED
PUBLISHERS

Frances Lincoln Limited
4 Torriano Mews
Torriano Avenue
London NW5 2RZ

www.franceslincoln.com

First Frances Lincoln edition: 2009

A catalogue record for this book is available from the British Library.
978-0-7112-3065-1

Printed and bound in China
9 8 7 6 5 4 3 2 1

Design: www.williamhall.co.uk

Throughout the book, the modern-day village of Downe in Kent has been called Down. (As Bernard Darwin put it: 'We always spelt it Down and thought for some mysterious reason that the final 'e' was low and modern.')

Illustration credits—
p4, p99 © Darwin Archive, reproduced by kind permission of the Syndics of Cambridge University Library
p14, p126–7 © William Hall 2009
p19 Image by courtesy of the Wedgwood Museum Trust, Barlaston, Staffordshire
p25 © Mary Evans Picture Library 2007
p57 © Tate, London 2008

Chapter quotes—
p12 Darwin's *Autobiography*
p38 Darwin's notes, Box B, University Library Cambridge
p66 Notebook C, 1838
p90 *The Descent of Man*, 1871, 'Mental Powers'
p118 Chapter 5: Notebook B, 1837-8

Cover animals, from left—
Pan troglodytes (chimpanzee), Australopithecus afarensis, Homo neaderthalensis, Homo erectus, Homo sapiens (human), Canis familiaris (dog)

For more images, ideas and dogs visit www.darwinsdogs.com

Darwin's Dogs

It is a summer's day in the early 1860s. The family of Charles Darwin, the notable naturalist and author of *The Origin of Species*, pose for a photo with their pet dog. The photo is framed around a window facing onto the lawn at their home in Down, Kent. Mrs Emma Darwin, mother of a splendid brood of almost grown-up children, sits in the window frame, wearing a bonnet and reading a book. On the far left of the image sits Leonard, a tall teenager of about thirteen in a jaunty cap. Next is Henrietta, the daughter who assisted her father with his work, standing underneath a parasol. On the windowsill with his mother is Horace, who was only twelve or so; he would later found a world-famous company making scientific instruments. Sitting in wide skirts and a funny awkward little hat that hints at the trickiness of her character is Elizabeth, known as Bessy, who was around sixteen at the time of the photo. And standing to the right of the window is Francis, just a year younger than Bessy, who would share most fully in his father's scientific work, and inherit most strongly the family passion for dogs.

There is one other person in the photograph: seated on the ground is an unidentified visitor, given a less prominent place in the photo than the dog. We won't ever know the name of the visitor who came to Kent for one sunny day in around 1863, but we suspect the dog is Bob, a big black and white retriever who was made famous forever when his master described his 'hothouse face' in one of his last books, *The Expression of the Emotions in Man and Animals* (1872).

A 'hothouse face' is one familiar to every dog owner. It's a disappointed, hurt expression, with ears forward, full of pleading and the last traces of hope. In Bob's case, the disappointment concerned the missed chance of a walk. When Darwin left the house by the lawn door, Bob always believed they were both heading off down the garden for the morning's constitutional. He was excited. But if Darwin was actually going off to work in his little conservatory where he did his plant experiments, Bob was prone to take up a 'dejected attitude' at the turning in the path.

For Darwin, the hothouse face almost made him not want to leave Bob, and to carry on walking. But Darwin didn't accuse the dog of doing anything calculated: 'It cannot be supposed that he knew that I should understand his expression, and that he could thus soften my heart and make me give up visiting the hothouse.' For Darwin, the dog was simply acting on its instinct, trying to change his owner's behaviour, to make him leave work and go for the desired walk. Even in a simple interaction between owner and dog, Darwin analysed and noted, fascinated by the animal right in front of him.

Darwin recorded his dogs in his books and letters, though rarely as memorably as in the case of Bob. We have few photos of Darwin's dogs, loved as they were, because the Darwin family lived just on the cusp of photography becoming available to all. Darwin always consoled himself that he'd gone to the effort of having a glass daguerrotype taken of his daughter Annie, in a special trip to London two years before she died in 1851. Yet a mere decade later, a photographer could pack up their portable equipment and come to the house to record all the family, including Bob.

Though Victorian professional photographers advanced their skills with remarkable rapidity, the average black and white image still took many seconds to form, even outdoors on a bright sunny day. The dog in the photo sits at the family's feet, with his head down as if he's been tied to the floor. Perhaps he was tied: Bob has none of the ghostly white traces around his head indicating that he wriggled while the photographer exposed his plate. He seems to have lain perfectly calmly. But organising the photo must have been at least a little complicated nonetheless; perhaps that's the reason for the half-smiles on the faces of Darwin's wife Emma, and daughter Elizabeth.

The taking of a photograph perfectly sums up the huge gap between human being and animal. Dogs live within our households as members of the family. But whilst even quite small children can have the need to sit still for a short time explained to them, a dog must be ordered to stay. A dog cannot be persuaded by a bribe, or reasoned with using logic; only a direct order keeps a dog in place. Neither conversation nor persuasion are ever possible. There is an enormous divide between the human and animal worlds.

If the taking of a simple photo is subject to such a communication gap, you'd think that humans and dogs getting along in everyday life would be a complicated and chancy business, full of risks. Dogs are, after all, descended from wild animals who hunted for their survival. Human beings living with animals? The analytic intellectual ape, side-by-side with the unpredictable, inexplicable dog mind? Surely the lack of verbal communication, and the fact of two minds so different in so many ways, would render the relationship at least, well, tricky?

Yet for many, the relationship between human being and dog is the calmest, happiest one in their life. The least demanding, and often the most quietly rewarding. Globally, more than 200 million human beings living on the planet today choose to invite one of these animals into their lives: living with them in their homes, sleeping with them in their bedrooms, sharing with them their own food, leaving them alone with their newborn babies. And Darwin was one of them.

When we think of Charles Darwin's work on the animal kingdom, we might think of him handling the tiny bodies of Galapagos finches, or examining the massive shells of those island's native tortoises. We think of exotic animals, like gorillas and other monkeys, in the jungles of Africa or Papua New Guinea, thousands of miles from England. Yet the animals with which Darwin had the most profound and sustained contact were the ones that lived with him in his home. Throughout his childhood, and for the years of his grown-up life with his own family at Down House, Darwin owned dogs.

His life with dogs began with Shelah, Spark and Czar, the three most-loved dogs of his teenage years. Then at Cambridge University, he hunted with his cousin, William D. Fox, taking along their dogs Sappho, Fan and Dash. Next came another hunting dog, Pincher, and the little dog Nina, both left behind when Darwin set off for his five-year voyage on the HMS *Beagle.*

When Darwin had children, he also acquired dogs. Bob was the big piebald dog in the photo, a proper family dog, loved by all. Bran was a deerhound puppy, who arrived in 1870. The family were particularly good at adopting dogs; Quiz, Tartar, Pepper and Butterton came to them that way. So did Tony, who originally belonged to Sarah Wedgwood, Darwin's sister-in-law; Tony was taken in by Darwin after she died in 1880. And finally, along came Polly, the last dog Darwin ever owned; at first his daughter Henrietta's, Darwin held onto Polly after Henrietta married and moved from Down; she was the dog his son Francis said his father loved the most.

Dogs were the animals Darwin observed the most closely and for the longest; over his whole lifetime, except for his journey on the *Beagle,* Darwin spent almost every day in the company of dogs. Darwin grew up in Shrewsbury, a farming town, with cattle markets and agricultural fairs regular events in the calendar. He petted and walked with his dogs, watched them hunting, enjoyed wandering through the countryside with them.

But Shelah, Spark, Czar, Sappho, Dash, Pincher, Nina, Bob, Tartar, Quiz, Bran, Tony, and Polly were also some of the most important characters in the story of his thinking. He wondered what thoughts they had, he tried to explain their behaviour, he wrote letters to other people on the subject. This was not just idle speculation. Dogs stimulated his scientific thinking in a number of ways. And when he began seriously to consider a theory of evolution, he began his writing, not with finches or tortoises, but with domesticated animals like pigeons, cattle, poultry and dogs.

So when it finally came to the publication of *The Origin of Species,* that nerve-wracking moment, Darwin wove a peaceful sense of calm into his first chapter by starting off with a discussion of plants and animals straight from the English country farmyard. Ducks waddled, cows were milked, ears of wheat ripened: the whole agricultural year turned in the course of chapter one. Darwin used plant breeders and livestock experts to show that his theory of natural selection worked on the same lines as a picky dog breeder would, selecting for desirable traits, and eliminating harmful ones. Familiarising and domesticating this strange new theory, he grounded the whole story at home, ensuring it was clearly accessible for the general Victorian reader and making the bitter pill easier to swallow. Darwin's dogs brought evolutionary theory right to the hearth rug of the Victorian home.

And it was indeed a bitter pill. By the time the photo including

Bob was taken, controversy over Darwin's theory had risen to its full height. Reviewers became mudslingers as scholarly debate descended into angry name-calling. The strongest invective was reserved for the question of man's place in nature. If Darwin's evolution did happen, surely that must include human beings? Which meant that human beings were simply animals: no better, no worse. Many people were horrified by this notion: you didn't have to be a devout Christian to feel that human beings had special qualities that let them rise above the herd. Cooperation, altruism and religious belief were all pointed to as evidence that human beings were special. Special enough that these unique qualities must have been put there by a creative deity, rather than having evolved. What possible room, these critics asked, could there be for unselfish kindness, in Darwin's struggle for survival?

Nonetheless, advocates of Darwin's theory fought their corner fiercely. Sometimes it came down to fighting over the very microscope slide itself. Thomas Huxley, often called Darwin's 'bulldog', accused the top anatomist Richard Owen of having failed to look at his own dissections of gorilla brains properly. Richard Owen, already incensed, fell into Huxley's trap. His reputation was permanently tarnished because he had misinterpreted the brain anatomy of apes in trying to prove that human brains were unique.

But Darwin himself held off from becoming involved. He spent the ten years after the *Origin* composing his thoughts on the question. Darwin believed that human beings had evolved; but most of all, he believed that we shouldn't be insulted, or feel that we are being dragged down, when we accept our animal ancestry. And every one of those dogs contributed to his thinking. Every single one helped fuel his argument that what we might see as a huge gap, between human and animal, is actually less than we think. When we become angry, we can see the same behaviour in a dog. When we love, said Darwin, we can see the same behaviour in a dog. When we dream, we can look at a dog twitching and yapping whilst it sleeps, and know that they are dreaming too.

In one of his final books, *The Expression of the Emotions in Man and Animals,* he told the story of Polly, the white terrier who was his final dog. Polly was the one creature who sat with Darwin all day long as he worked in his study. 'He was delightfully tender to Polly,' wrote his son Francis in a memoir of his father, 'and never showed any impatience at the attentions she required, such as to be let in at the door, or out at the verandah window, to bark at "naughty people", a self-imposed duty she much enjoyed.' As Darwin aged, Polly's dog basket stayed firmly by the fire in his workroom. You can see the basket in photos of the study taken by the family. In the end, Polly lasted only a few days longer than her master; she was put down by Francis, and buried in the garden. Darwin lived his whole life as a dog person. So let me invite you now to a rather different account of the life of Darwin, this one told from the canine point of view. Others have told the story of the finches and the tortoises; now let's hear the tale of Bob and Polly.

To my deep mortification my father once said to me, 'You care for nothing but shooting, dogs, and rat-catching, and you will be a disgrace to yourself and all your family.'

Charles Darwin grew up with dogs. We know this because in a world where communication was slow, the Darwin family wrote letters. Letters were written describing family parties the boys had missed; the boys wrote requiring things from home to be despatched to needy boarders; and letters came back by return from their older sisters, exchanging news and private family jokes. What we know about the exact day-to-day flavour of the childhood of Charles Darwin, we know from these letters.

The letters show that the Darwin dogs were intimate members of the household. We don't have images of these early Darwin family dogs, for they lived before the advent of photography and were never painted. But we know a little about their appearance and temperaments. Shelah was an all-round family dog: Dr Darwin's daughters would later keep one of her puppies as a pet, too. Spark had a fierier temper, a much-loved black and white mongrel who was a distinguished rat-catcher. And Czar was a particularly aggressive dog, who was eventually banished for bad behaviour.

And where Charles was concerned, letters and private family jokes often centred on these dogs. Dogs are on occasion referred to with humorous formality as 'Mr' or 'Mrs': in 1826, Darwin's sister Caroline wrote to him: 'Mrs Shelah condescends to pay me much more attention than when you were at home; she does not get much exercise beyond her daily walk into town and a little romping with any odd apple which she entreats me to throw down the bank for her to pick up.' More frequently, the dogs are talked about as children; when Darwin's sisters wrote to him, Spark was 'your child', and once even 'your favourite child.' The story of the Darwin family was closely interwoven with the story of their dogs.

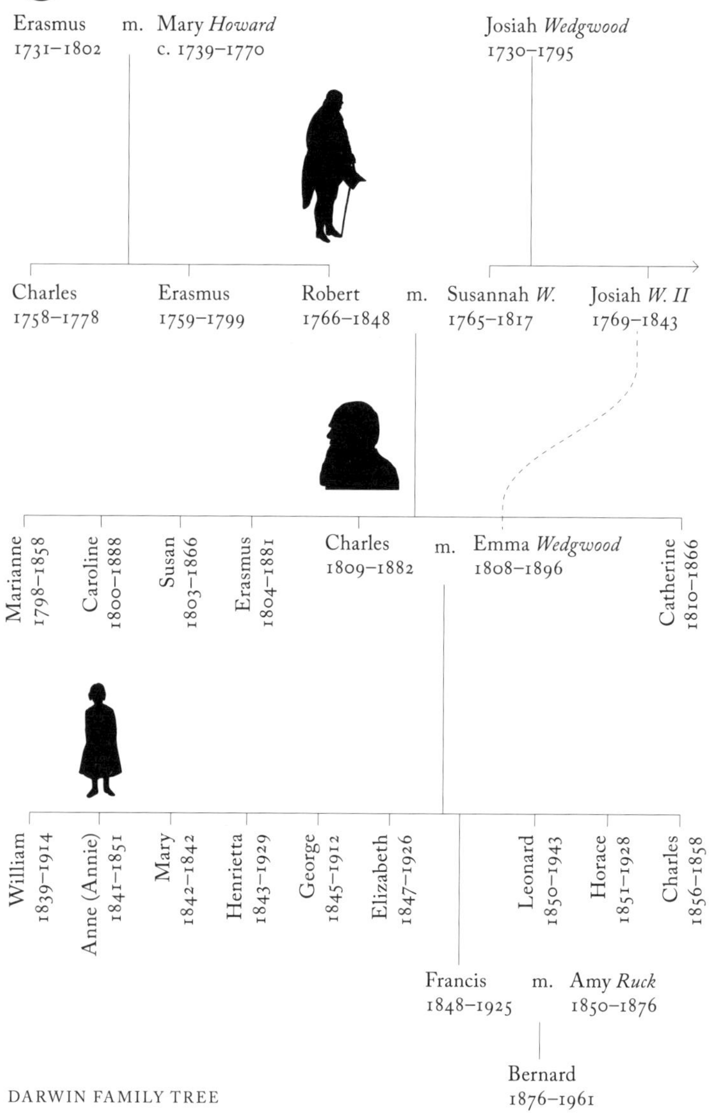

DARWIN FAMILY TREE

Charles was born on 12 February 1809, in Shrewsbury. His mother Susannah was forty-four and already had a substantial brood of children, overseen by the towering figure of their father, Dr Robert Darwin. Charles had three doting older sisters, Marianne, Caroline and Susan, and one older brother, Erasmus. The three girls came first, a natural trio; then came the two boys. And after Charles, one more baby was born; the doctor and his good wife added a little sister for everyone, Catherine.

Predictably, the two boys were easy companions, both out in the countryside exploring, then at school. Charles was five years younger than Erasmus, and had to work hard to keep up with him. Their whole educational careers were spent together, with Erasmus always forging ahead, relaying his excitement over what he was learning back to his younger brother. And Charles was perpetually hurrying to find something of his own to impress his heroic elder brother, who always got to know everything first.

Charles and Erasmus's early education took place at home, at the hands of their mother and sisters. Then, later, they were sent to a little day school supervised by a Unitarian minister, Unitarianism being their mother Susannah's preferred strain of religious belief. And after the boys lost their mother in 1817 they were sent to boarding school together just a mile from home in the town of Shrewsbury. Darwin was just nine years old, Erasmus was fourteen, but the family trusted that Erasmus would keep an eye on his much younger brother, which he did.

From their early years, we have few letters. But in 1822 at the age of eighteen, Erasmus went off to Cambridge to follow a medical course. Letters came home in a flurry, asking for various forgotten books to be sent to him, and requesting money and experimental equipment ranging from pieces of platinum wire to ten pound notes. Darwin clearly longed to follow his brother; he set himself to reading the books that Erasmus had been assigned by the Cambridge lecturers, and arranged to go and stay at Cambridge in the summer holidays of 1823. The letters are busy and practical. Emotion is absent, articulated in one way only: affection is sent not to Charles, but to Charles's dog. When Erasmus writes home from Cambridge in 1825, his letter ends with the cheerful instruction to his younger brother to 'Greet Spark'.

The boys grew up quickly. In 1825, Erasmus was ready to begin his clinical training at Edinburgh University, and Dr Darwin decided that the brothers should no longer be separated. Medically trained himself, he took the decision to send both his sons to university in Edinburgh to follow in his footsteps. Edinburgh was one of the best places to study medicine in the world, and Dr Darwin had high hopes for both of his boys. Erasmus could do his clinical training, for which he'd qualified at Cambridge, and Charles could begin his medical studies: anatomy, chemistry and midwifery. Charles was only sixteen, but the Doctor hoped that Erasmus would set a good example. So despite Charles being relatively young, the boys were sent away together, with Charles, as always, trailing in Erasmus's impressive wake.

And so the letters went back and forth. The boys typically wrote short notes, reassuring their father and sisters that they had been to church, and omitting to report when they missed lectures. Longer letters came by return from Shrewsbury. Their elder sisters included helpful spelling corrections and reports of Shrewsbury social life, such as a week of theatre visits to see William Charles Macready,

OLD SHREWSBURY SCHOOL IN 1811

one of the period's most celebrated actors. There are glimpses of the family dynamics at work: Caroline tells a funny story about Susan, and Susan then writes crossly, 'Caroline's amusing little anecdote at my expense on the other page I beg to contradict as it is quite false.' There were plenty of jokes like this, all conducted with the family's wry sense of humour.

But the letters are also tender. Caroline wrote to Charles when he had only been a few weeks away at Edinburgh: 'I assure you I miss you very much ... and as for poor Shelah and Spark, they look the pictures of melancholy, and are really grateful for any notice, from anybody.' For the Darwins, talking about the dogs became a way of talking about their own feelings: the dogs were said to miss people, a way of expressing that sadness at somebody's absence. The dogs offered a way of speaking intimately to each other. Charles was a motherless teenager out in the world for the first time, and conveying news about the dogs was one way for his sisters to know they were talking about something he really cared about.

The household dogs may have been members of the family, but not all dogs were friends. When Darwin remembered his childhood much later in life, he recalled being very afraid of the unfamiliar dogs he met in Shrewsbury town, 'in Barker Street', he wrote, thirty years later still able to conjure up the very spot; looking back, he commented, 'I was very timid by nature'. And he also remembered vividly his own shame at acting badly towards an animal:

> Once as a very little boy, whilst at the day-school, or before that time, I acted cruelly, for I beat a puppy I believe, simply from enjoying the sense of power; but the beating could not have been severe, for the puppy did not howl, of which I feel sure as the spot was near to the house. This act lay heavily on my conscience, as is shown by my remembering the exact spot where the crime was committed. It probably lay all the heavier from my love of dogs being then, and for a long time afterwards, a passion.

Charles's sisters sometimes sent up that 'passion' for the dogs. Caroline wrote while Charles was first in Edinburgh: 'Spark we hear from Overton is in high preservation,' and then added, tongue firmly in cheek, 'and so is your little nephew who of course you have a much stronger affection for.' Darwin replied, in the same vein: 'I hope ... you will send me many more nice affecting letters about dear little black nose. Erasmus thinks I shall have more pleasure in seeing it than all the rest of the families put together. You seem to

THE WEDGWOOD FAMILY, STUBBS 1848

hold the same opinion with regard to my dear little nephew.'

This sentimentality towards the dogs is intriguing. Darwin's parents had both come from landed families – Dr Darwin from landowning stock, and his mother Susannah from the fashionable Wedgwood clan. Both families owned animals they were proud of, but the Wedgwoods were particularly keen to celebrate their animals. As a consequence, Susannah and her siblings had their family portrait painted with their horses, by no less an artist than George Stubbs. According to their unimpressed father, Josiah Wedgwood, the painter managed better likenesses of the horses than the children.

The Darwin home was a substantial Georgian house in town, but Dr Darwin's income came from other kinds of investments, including mortgages on agricultural land, so the house dogs were not the only domestic animals Charles knew as a child. Charles's very earliest memory was of a farmyard escapee, when he was less than four years old: 'I was sitting on Caroline's lap in the dining room, whilst she was cutting an orange for me,' when he saw 'a cow run by the window, which made me jump'.

The money for the Darwins' leisured lifestyle originally came from both sides: Susannah had come to the marriage with a nice inheritance from the Wedgwood potteries; Dr Darwin was a medical man by training, earning him a tidy income of several thousand pounds a year. However Darwin's father didn't settle for this. He spent some of every day overseeing his financial transactions; he used his land to mortgage and remortgage, not like a character in a Victorian novel, because of debt, but in search of a good interest rate. Dr Darwin lent money to local entrepreneurs who needed it, and waited for a return. He invested in all the symbols of the new industrial era: canals, toll roads and bridges. He also lent to lords, viscounts, schools, jails and waterworks. Wherever there was a return to be had, Dr Darwin would send his money.

Cows, pigs, horses and chickens would have wandered through Darwin's mental landscape, as familiar as the Shrewsbury scenery. But where the extreme tenderness towards the animals came from is an interesting question. In the early nineteenth century general attitudes towards animals became considerably more sentimental. And it is also possible that it is connected with the loss of Darwin's mother. Charles later wrote that he could recall hardly anything about her death, but that Catherine, a year younger, seemed to remember every detail. With four sisters to dote on him and an admirable older brother, he didn't lack attention and love. But he lacked a mother. In this jolly but motherless household, he formed the kind of intense bonds between boy and dog that are familiar throughout history.

A good example of that intensity comes from Darwin's teenage years. Up till now his passion for dogs had been mostly a thing for light humour. But the year he was seventeen, it became a matter of the greatest seriousness. In February 1826 his sister Marianne wrote from her home at Overton:

> It will sound odd to begin my letter with telling you that I never in my life was more sorry to write a letter than I am to write this one, but it is the case.

Marianne was newly married and was looking after Spark, Darwin's own dog, 'dear little black nose', for a time while Charles was away at university. Marianne had to break the news that Spark had died whilst in her care. Marianne was absolutely mortified and the letter she wrote is full of her distress.

> I believe you know that we begged to have poor little Spark for a short time 'till we found another, but I don't think you were told that the day after she came to us, she ran away, and though we made every possible enquiry and search, we heard nothing of her for a fortnight, when we found her established at a gentleman's house in this neighbourhood and the grand pet of all the family. We got her back again, but we soon perceived that she had during her absence from us become with pup. We were very sorry for it, as we knew you did not wish it. Last Monday the poor little thing was taken ill, and after the birth of one puppy she died. You cannot think how sorry we have all been about

FOX TERRIER.

> it. Everybody in the house had got so fond of her, and she was such a nice little dog. I hope you will write to me my dear Charles – for though I have been very sorry for poor Spark's death on her own account, I have been still more so on yours, and altogether it has vexed me more than any thing that has happened for a long time. Shelah is going to have a family, but I am afraid they will not replace poor little black nose.

We needn't doubt Marianne's sincerity; she had never felt more discomfort about breaking news. Her teenage brother Charles, away from home properly for the first time, now had to deal with the loss of 'dear little black nose', his favourite pet.

We don't have Charles's reply to his sister, but he obviously wrote with some feeling: she later talked of how 'severe' his return letter had been. Another sister, Caroline, also wrote to Charles from Shrewsbury, feeling the same difficulty in knowing what to say to her brother. 'I am sure poor Spark's death will be a sad grief to you, independent of the sorrow for the little dog herself. I do not think you will understand me but I do not know how to express myself clearly'.

Caroline didn't often give up on trying to express herself; the level of feeling suggested is deep and sincere. Finally, Marianne wrote sadly: 'We are so unlucky in puppies, that I hope we never shall have any more … I hope to hear from you again, my dear Charles, as I should be very sorry if our correspondence stops here.'

Marianne had offered to compensate Charles by giving him her own new puppy, but there was thankfully some good news. Shelah was pregnant too; so instead of taking the Overton pup, at Shrewsbury they waited for Shelah's litter, due to be born any day. Soon, Caroline and Susan had a new pup to write to Charles about, who within weeks was 'so fat, he can neither stand or go,' wrote Susan, as sharp-witted as ever.

Their little sister Catherine was not much of a correspondent, but even she pressed herself into writing to laugh at the newcomer: 'Shelah's puppy is really quite a sight to be seen; it is as broad as it is long.' After the episode with Spark, there were to be no more accidents and the girls were all now fiercely protective of the new baby animal: 'We had an alarm that he was poisoned last week by some paint, as the house and premises are infested with a troop of little painters with Mr Pierce at the head of them, who are painting the windows etc etc. Caroline scolded all the painters round and afterwards found her alarm was false.'

All the family were protective, but the dogs weren't completely indulged; some rules were observed. One dog, Czar, was removed from the household permanently for biting. Yet in general the Darwins were very soft-hearted about their pets. Nina, a little dog who belonged to Charles, was left in the family care while he voyaged round the world on the *Beagle*. In late summer 1832 she was attacked by a horse who bit her leg, lifting her up and refusing to put her down. 'Her leg was badly broken,' wrote Caroline anxiously.

But rather than have the suffering dog put down, the Darwins ensured she was 'attended' by 'surgeons'. Caroline wrote straightaway to Charles, oceans away in Valparaiso, Chile, to allay his fears: 'she is getting well and does not seem to suffer any pain now.' The latest news about Nina was considered important enough to be reported before the news that cholera had come to Shrewsbury and that many people had died: 'it is now some days since we have heard of any death', wrote Caroline, almost as an afterthought to Nina's leg.

Gossip about other people involved dog talk too: in January 1826 Caroline recounted in a cheerful letter to Darwin how she had met the 'quite exquisite' Mr Gibbon, in an episode out of the pages of Jane Austen; 'so handsome, and so conscious of it that he could not speak or turn his head, without thinking he was a study for a painter and model to a sculptor'. The self-consciously handsome young man tried to charm Caroline by telling a story about a young lady rescued from drowning in the sea by a Newfoundland dog, adding finally 'I always call that dog a gallant fellow.' Caroline commented, drily, 'I give this as a specimen of the good taste of his conversation.'

NEWFOUNDLAND DOG SAVING A CHILD, J ROGERS, AFTER BEAUME

It's not surprising that the handsome Mr Gibbon should have tried to charm a lady with such a story, for the heroism of dogs was a subject of profound interest to a sentimental early nineteenth-century social gathering. Prints showing 'A Newfoundland Dog Saving a Child' had been a popular moral subject in the Romantic household, and were a common sight on the walls of Georgian houses.

The subject became even more widely loved in the 1830s, when Edwin Landseer produced several paintings of Newfoundlands rescuing human beings from the water for a new generation of sentimental beings. These were reproduced as prints, leading to huge sales and some steamy copyright battles. Landseer was immortalised for his efforts: 'Landseer Newfoundland' is still the name given to the black and white colouring in that breed of dog.

The idea of the noble dog rescuer stayed popular throughout the century, and in Victorian times the image made its way into children's books, and onto mahogany panels, postage stamps and other household items; in Henry James's *The Bostonians* of 1886, Mrs Luna's boarding house even has a rug showing a Newfoundland saving a child from drowning.

The Victorians were fascinated by the morality of the animal kingdom; children were encouraged to read books where human-sounding bees taught lessons about working for the greater good of society, and about hard work and sacrifice. Dogs that risked their lives to rescue a child epitomised these stout Victorian values. But the period also saw growing sentimentality towards animals. More compassionate legislation was one result: in 1822, Richard Martin saw the first anti-cruelty bill through Parliament, which helped lead to the founding of the RSPCA in 1824.

A debate about working dogs followed, in the early years of Victoria's reign. In the little book *Dogs and Cats and How to Manage Them,* published 1882, the author wrote about how in his youth he was used to seeing 'dogs harnessed to carts belonging to bakers, butchers, cat's-meat vendors, and costermongers of all trades.' Dogs were cheaper and more obedient than donkeys; yet, the author argued, the Act of Parliament meant to rescue them from ill-treatment led to thousands of working dogs being drowned, as owners could no longer afford to feed them. 'Within a month, there was scarcely one of these useful dogs to be seen in the streets of London.' Victorian attitudes to dogs were complicated, and not always compassionate.

The loyalty of dogs was also much-discussed. This, after all, was the era when Greyfriars Bobby became famous for his punctilious attendance at the grave for fourteen years after his master's burial. But the Transcendentalist American writer Henry David Thoreau was more sceptical about attributing high-minded motives to the canine race, wrily questioning the idea that a dog is morally good because of these sorts of actions: 'A man is not a good man to me because he will feed me if I should be starving, or warm me if I should be freezing, or pull me out of a ditch if I should ever fall into one. I can find you a Newfoundland dog that will do as much.'

But many years later, when Darwin was in his sixties and writing *The Descent of Man,* he seems to have remembered the example of that brave Newfoundland dog. In chapter four, on 'Moral Sense', Darwin describes dogs that are altruistic enough to attempt to save the lives of individuals belonging to another species. The behaviour of a Newfoundland dog that rescues a human being is the central example for most Victorian commentators on the nobility of dogs, and Darwin was fascinated by this. He sought to understand just what it was that a dog was doing when it rescued someone, when it remembered someone or when it conceived a lasting hatred for someone. All this would form part of his argument that animals had many qualities which had previously been thought to be unique to human beings. If dogs, said Darwin, can be capable of love, of hatred, and of altruism, how unique can we really be?

It wasn't just Charles who was interested in the way that dogs behave. Before Spark's death, Susan and Caroline had gone to visit Marianne at Overton. Caroline sent Charles a few sweet lines about Spark. But Susan reported on the dog with a more detached eye and a scientific approach to detail. Susan was interested in her brother's assertion that Spark would remember him, despite their having being separated for some time. Susan took a certain amount of contrary pleasure in reporting that Spark had 'growled and snarled' at the two sisters, even having the 'impudence' to snap at Caroline's finger.

However, reported Susan, another little experiment on Spark had more intriguing results: 'The name of "Shelah, Shelah" produced a very visible effect on her, as she pricked up her ears, and looked excessively puzzled.' Susan was interested to report this because of family debates about whether the dogs would remember them after an interval. It was clearly not just Charles who was intrigued by the intelligence and abilities of mute animals: the letters show that the family had debated these questions together. And Charles's interest in whether the animals remembered him continued when he set off on the voyage of the *Beagle,* as his sisters' letters reveal.

Darwin's interest in his dogs tells us something important about him as a man. It tells us that he had strong and personal relationships with these animals; that he saw them as playing a central part in his life. Dogs were the animal that Darwin had the greatest opportunity to observe. And later on, when Darwin came to believe that living things were related in a single huge family tree, it would become clear that stating something about a dog would mean that it was likely to be true for animals higher up the tree also. But even before Darwin began his theorising, amongst the family there was clearly a tradition of thinking about dogs, pondering what their characters meant and speculating about what their behaviour demonstrated. You might even conclude that growing up in this kind of a family helped to stimulate Darwin's own thinking about the relationship between human beings and other animals.

*   *   *

Each summer holiday, returning from Edinburgh, Darwin moved through an agricultural landscape. England was changing, and shifts in farming practice were making their mark on the countryside, just as the rapid growth of industry was changing the towns. Land was being enclosed, crops were being rotated, livestock was being bred for specialist purposes. New machinery for labour-intensive jobs like threshing also played its part. Farmers could now grow more food with fewer workers; the laid-off agricultural labourers had little choice but to find jobs in towns and cities.

In this climate, farmers wanted proven information about crops, fertilisers and cultivation to continue with their 'improvements'.

Agricultural societies were an important channel of information, encouraging farmers to share experience and knowledge about systems and practices. The Bath and West of England Society is a famous example, established in 1777 by a group of philanthropists led by Edmund Rack, a draper who believed in improvement of agricultural practice. Weekly meetings, annual shows and learned journals helped to create an arena for the discussion of agricultural matters. The circulation of agricultural journals and newspapers also rapidly grew, as farmers became aware of the possible value of new ideas. The best way to fence stock securely, the best way to fumigate for pests: all innovations were discussed in search of the greatest productivity.

Even Shrewsbury was changing. Though a medieval town, it sat in the shadow of the Potteries, and going on his father's medical rounds with him, Charles cannot have escaped seeing the looms in the corners of rooms where working people did piecework to make up their meagre wages. He often took notes for his father: when he was sixteen he spent the summer holidays riding round with his father treating poor patients, mostly women and children in Shrewsbury. He quickly acquired patients of his own, despite his age. Most of these were the urban poor, affected by all the new change.

Charles would have seen the countryside, too, chatting to farmers as he crossed their land, as he went hunting. He was living amongst people preoccupied with the new questions of how best to make land productive, and how to look after livestock and crops to ensure the best return. The Darwins were an early Victorian family, and being around livestock was part of their existence. So it was in this firmly agricultural setting that Darwin would first have encountered the question of inheritance; not with respect to foreign exotics, but in connection with farmyard animals, such as cattle, sheep and dogs.

Farmers were particularly keen to have good information about breeding. Though the agricultural world was full of skilful breeders, there was little understanding of why certain techniques worked. This didn't hold back the improvement of breeds: cows, pigs and sheep were subject to intense work during the nineteenth century. Friesian herds were developed in Northern England; Pembroke cattle in Wales; Aberdeen Angus in Scotland. And in 1822, the Shorthorn of Durham became the first breed of cattle to have their own pedigree herd book, which recorded the family tree of every cow considered to be 'of the breed'.

Breeders of dogs engaged in similarly enthusiastic exchange of information. Encyclopedias of 'Rural Sports' gave definitions of breeds; articles now appeared in *The Farrier* and *The Veterinary.* Until the mid-nineteenth century, dogs had only belonged to a few recognisable breeds. For example, in Bell's *British Quadrupeds* of 1837, the author lists less than twenty breeds of dog, including bulldog, greyhound, terrier, dalmatian and spaniel.

Dog breeders in the past had simply bred the kind of dog they needed, without reference to anyone else's criteria. Some dogs' only purpose was to keep an owner company, others were for hunting, for ratting, for flushing and retrieving, and for guarding and herding livestock. Breeders kept the dog which best performed the task, and bred from that individual.

But in the mid-nineteenth century, dog fanciers became increasingly preoccupied with establishing breed standards, and with safeguarding the purity of bloodlines. In the 1830s and 1840s, breeders began to think about how to perfect breeds, and the keeping of stock books to record pedigrees became common practice. In 1859, the first organised dog show was held, and in 1873, the Kennel Club was founded. These events marked the formalisation of those existing loosely-defined breeds, such as the greyhound, but it also witnessed the proliferation of new breeds, such as the Sealyham terrier and the Dandie Dinmont, named after a character in a fashionable Victorian novel by Walter Scott.

So many dogs, with so many different attributes; but where did all this splendid variety come from? 'Who can believe,' Darwin later wrote, 'that animals closely resembling the Italian greyhound, the bloodhound, the bulldog, or Blenheim spaniel, ever existed freely in a state of nature?' We don't know when the question first occurred to the curious young man about the 'natural state' of dogs. But later, when he himself became interested in the question of breeding and inheritance, it was to livestock experts and seed breeders that he would turn. When Darwin wanted to know exactly how particular breeds had come into being, and how particular traits were bred into an individual, he began with these men, ordinary men whose richness of experience offered him a body of knowledge to begin mining for general principles.

But before that, he had to finish his education. Thanks to his passion for shooting, dogs and rat-catching, Darwin rather distracted himself from orthodox studying during his university years. Whilst he worked hard to become expert in the classification of tiny sea creatures, he was less attentive to the medical curriculum he had been sent to Scotland to study. And at the age of eighteen, after two years in Edinburgh, all his most exhilarating moments happened

whilst out hunting. During his university days, he admitted in his autobiography, he even justified going out with the dogs because of the skill involved.

> How I did enjoy shooting, but I think that I must have been half-consciously ashamed of my zeal, for I tried to persuade myself that shooting was almost an intellectual employment; it required so much skill to judge where to find most game and to hunt the dogs well.

Dr Darwin, Charles's stern father, eventually conceded that perhaps Edinburgh and medicine had not been the right choice. But after a switch to Cambridge, the hunting continued on an even larger scale, as Darwin fell in with a crowd of what he later politely termed 'some dissipated low-minded young men.' Still, he confessed, he found it difficult to look back without remembering what fun it had all been.

THE POINTER.

Dogs were a central part of the fun. Darwin borrowed other people's dogs and wrote enthusiastically of dogs to the other dissipated low-minded young men. The overall goal was one they all shared: to go out for the entire day, trampling through the countryside, roaming unfettered, enjoying the seasons, admiring the world. From his cousin William Fox, he took a gundog called Dash, whom he liked so much he awarded him the family honorific 'Mister':

> I and Mr Dash arrived quite safe here on Saturday morning. He rises in my opinion hourly, and I would not sell him for a five pound note. It would have excited your envy and spleen to have seen him on the scent of a covey of birds, and the style in which he went down when I held up my hand.

Darwin was strict with the gundogs, though. As a student, Darwin expected his hunting dogs to be obedient, never chasing game without permission. When he was away with the *Beagle,* Caroline wrote, recounting a country walk: 'Pincher still remembers your training so well, that though a hare sprung up just before us and he looked as if he would have given worlds to follow, he obeyed and walked close behind me without attempting to have a hunt.'

And despite their worth as hunters, these dogs also were the subject of sentimentality. When Darwin was away on the *Beagle,* his sister Caroline spelt out quite explicitly the relationship the Darwins had with the dogs: 'I have no more family news, except poor Pincher has cut the sinew of his foot with a glass bottle and they fear will be lame for life'. Pincher's accident counted as important family news, in the same way as a new baby or a marriage might.

When Darwin needed information on the breeding of dogs, a few years down the line, Dash's owner William D. Fox was one of those who Darwin quizzed. Fox was a second cousin; his grandfather was the older brother of Erasmus, Charles's grandfather (the D stood for Darwin). Charles sent many letters picking Fox's brains, hoping to make use of his canine expertise. Fox always remained Darwin's friend, but he also became part of the team of experts providing evidence Darwin needed to shape his theories. Whilst at Cambridge, though, the relationship was simply about the shared enjoyment of the countryside on a beautiful autumnal morning.

Darwin's academic career ended up rather chequered, though it is unclear whether the dissipated young men and the passion for dogs and hunting were completely to blame. Having broken the news to his father that he was unable to stomach the idea of becoming a doctor (unable, rather literally, for watching operations made him feel sick) they agreed he should go for a degree at Cambridge, intended to enable him to go into the Church. Darwin slightly slouched through the course, but became extremely expert and keen on beetle collecting and geologizing – two subjects unfortunately not featured on the examination syllabus.

As a slightly absent-minded, easy-going younger brother, with a hard-working father and a bright older brother, Charles grew up in the shadow of the achievement of others. Perhaps it is not too surprising that he always felt his father judged him rather too dreamy and placid, and frustratingly lacking in the drive that had made the paternal fortune.

Understanding Darwin's central place in our own world, it's hard for us to imagine now how Dr Darwin could have feared his younger son would come to so little: hunting, dogs and rat-catching. Yet it would be a long time before Charles would feel his father had confidence in him, and that his father no longer worried Charles would be 'a disgrace to himself and all the family'. For Darwin, his interest in dogs was a symbol for his father of his son's lack of real direction. Darwin always felt his father's anxiety; yet his sister Caroline, on the other hand, thought that Darwin had always read his father wrong: 'Charles does not seem to have known half how much my father loved him.'

After finishing at Cambridge, Darwin returned home still very unsure as to which career to follow. He could imagine just about managing to become a country parson, yet the idea failed to arouse any of the passion he felt for study of the natural world. His salvation came in the form of his former botany tutor back in Cambridge, John Stevens Henslow, who found him a position as companion to Robert Fitzroy, captain of the HMS *Beagle,* about to set off on a two-year surveying voyage round the coasts of South America. It would be five years before they would return to English shores.

Even at the moment of leaving for the voyage, he saw things in a distinctly domestic way: 'it strikes me,' he wrote before leaving, 'that all our knowledge about the structure of our Earth is very much like what an old hen would know of the hundred-acre field in a corner of which she is scratching.' Determined not to end up an 'old hen', Darwin set off around the world.

One out of every hundred litters is born with long legs, and in the Malthusian rush for life, only two of them live to breed. If prey are swift the long-legged one shall rather oftener survive... in ten thousand years the long-legged race will get the upper hand.

It was 2 October 1836 when the *Beagle* sailed back into Falmouth Harbour. Five years had passed since the ship's crew had last been at home, and Darwin's first sight of England would have been the shores of Cornwall, just glimpsable through the evening darkness. He had been to Rio, Tasmania, Tahiti and Hawaii; he had stood on the wild shores of Tierra del Fuego; crossed the plains of Patagonia on horseback; watched the Chilean volcano Osorno grumbling and smoking; and had ambled amongst the gum trees of Australia's Blue Mountains. But what Darwin wanted to do right now was get home to Shrewsbury.

He wasted no time sending word ahead to warn of his arrival; he jumped on the first coach, travelling for two days non-stop, gazing out of the window in a state of renewed wonder at the beauty of England's green landscape. He planned to turn up at home unannounced as if it were the most usual thing in the world, and must have bounced along on his journey smiling to himself at his plan.

He finally arrived in Shrewsbury late on the Tuesday night, after everybody had gone to bed. Anticipating the delight of a good trick, he went straight to bed without waking anyone, planning to surprise his sisters the next morning at the breakfast table. And clearly he did surprise them. We can only imagine the shrieks of delight when they saw their brother after five years: the reunion was joyful. The whole day became an unplanned party; the servants all got drunk. 'I am so very happy,' Darwin wrote to his uncle Josiah.

But even during this first day at home, Darwin had dogs on his mind. 'I wonder if Pincher will be very glad to see you again,' his sister Caroline had written, just after Christmas in 1833, whilst amusing herself in a letter imagining her brother riding around the South American pampas like a Gaucho. Both brother and sister were intrigued to see what would happen in reality when Darwin returned. So how then did Caroline's test of memory work on the dogs?

Darwin tried the experiment that very day. As Francis Darwin later wrote: 'My father had a surly dog, who was devoted to him, but unfriendly to everyone else, and when he came back from the *Beagle* voyage, the dog remembered him, but in a curious way, which my father was fond of telling.' Here, then, is Darwin telling the story:

> I had a dog who was savage and averse to all strangers, and I purposely tried his memory after an absence of five years and two days. I went near the stable where he lived, and shouted to him in my old manner; he showed no joy, but instantly followed me out walking, and obeyed me, exactly as if I had parted with him only half an hour before. A train of old associations, dormant during five years, had thus been instantaneously awakened in his mind.

The result of the memory experiment was clear. This dog who hated strangers, didn't so much as growl when Darwin called for him. He instantly remembered Darwin after five years: for a dog, almost half a lifetime. For Darwin, this was proof that the intellectual abilities of a dog are relatively complex.

We don't know the name of this 'savage' dog who nonetheless remembered Darwin. It seems unlikely to have been Pincher, the obedient hunting dog. It's possible that the savage dog was Czar,

THE BULL-DOG.

the dog who'd been sent away for biting. However we do know that the intermission of five years and two days described here, so modestly, is the enormous watershed of the voyage of the *Beagle,* and it means that Darwin went to find his dog for the experiment on his very first day home.

Darwin was happy to attribute a sophisticated emotional make-up to a dog, comprising a 'train' of old associations, giving the dog a long and effective memory. But having always been interested in dogs as individuals, Darwin was now poised to look at the dog in terms of its wider relationships, as he embarked on a process of examining the connections between all living things. A process that in the end would lead him to the surprising conclusion that dogs and human beings are closely related animals, with a remarkable amount in common.

Darwin was now back in England to stay. He began to play a part in London scientific life, going to meetings at the Geological Society and the Geographical Society, and making connections with distinguished colleagues like Charles Lyell, the geologist. Dr Darwin gave his son an allowance of four hundred pounds a year, allowing him the freedom to work without worrying about money. Charles's only complaint was that he missed the countryside: 'I do hate the streets of London,' he wrote, grumpily.

Darwin's biggest task was to find experts who could help him classify all the material he'd collected on the voyage of the *Beagle,* which ranged from fungi preserved in glass spirit jars, to huge prehistoric bones. For the birds, which were preserved as sets of skins and feathers, he found a man called John Gould. Gould's father had been a gardener, and John also worked for his living, stuffing, mounting and classifying specimens at the Zoological Society for a hundred pounds a year. Despite his lowly status, though, he was extremely erudite and took the birds on willingly, to work on in his spare time. The enormous ancient bones, on the other hand, went to Richard Owen, a young anatomy hotshot at the College of Surgeons, who quickly declared them to be giant extinct llamas and capybaras; they were closely related to the species that still roamed South America, but far larger in size. Owen's findings intrigued Darwin greatly.

Gould also came back to Darwin with results. Gould explained that despite their huge diversity, the Galapagos birds were all closely related species. Gould's careful study of the little bird skins established that the Galapagos specimens were all finches, 'an entirely new group' of thirteen species. Each finch looked so different, Darwin began to think, because each came from a different island or islands with different living conditions. Wanting to know for sure, he found he'd forgotten to attach labels saying which island each was from; he ended up scrabbling to track down specimens shot by other crew members to reconstruct his own collecting trail.

Darwin was doing two things at once now. In his public life, he was organising experts for the specimens and hurrying to finish writing his account of the voyage of the *Beagle,* which he was getting ready to publish. But in his private life, conducted in his personal notebooks in scrawled writing he can't have imagined anyone else would read, he was beginning to develop a theory.

Darwin was tackling what was called 'the species question'. In 1836, most educated people had been taught that species were 'fixed'. God had created the world as described in Genesis, and each original pair of living beings was made perfect and complete, two for each species on earth. To doubt the fixity of species was to doubt the perfection of creation itself. But by early Victorian times, it was not unknown to suggest that species changed over time. Any amateur South Coast fossil collector could see that there were traces in the rocks of creatures such as ammonites which no longer existed on earth. And for example even as far back as the eighteenth century, Erasmus Darwin, Charles's own grandfather, had proposed that all life came from a single ancestor.

So had Erasmus Darwin, all those years earlier, actually been correct? From the tiny creatures found fossil-hunting at the seaside, to the bones of huge extinct animals like the newly-named dinosaurs, the earth's treasures had many Victorian observers questioning whether life on earth was really so very unchanging. Especially when, like the South American bones Darwin sent to Owen, the extinct animals seemed to be closely related to a living species. Perhaps the extinct animals were the living species' ancestors?

This theory, called 'transmutationism', was especially popular amongst radical young men who embraced its promise of political and social change. What was lacking was a mechanism. Species might perhaps change, but until someone suggested a convincing account of how this strange process actually came about, educated opinion remained on the side of the 'fixed'.

Newly returned from the *Beagle* voyage, Darwin had time to reflect on all he had seen. What Darwin had found amongst Gould's finches was just one piece of evidence; overall, a pattern was emerging. What he saw in his collections from a global voyage turned his attention back to the domestic animals he'd grown up with. The degree of diversity he saw in the Galapagos birds reminded him of the huge spread of chicken varieties, of fancy pigeons, and of dog breeds he'd seen growing up in Shrewsbury. In particular, he wanted to know how living things produced such incredible variation – from the huge range of different Galapagos finches, to the toy poodles, bulldogs and bloodhounds of Great Britain, all bred from an original wild ancestor.

In a world where each species was seen as a perfect creation of God,

1. Esquimaux Dog. 2. *Canis Dingo*, the Dingo of Australia. 3. Mexican Lap-dog. 4. *Cuon primævus*, the Buansuah.

observers might conclude that an individual which differed slightly from others was a 'freak', flawed in some way from the original perfection of design. But what if the tiny difference actually gave that creature a slight advantage over others? As Darwin pointed out in the quote that heads this chapter, in any group of carnivores that had to chase fast-moving prey, the longest-legged, fastest runners would be first to catch their dinner.

Darwin was particularly interested in these kind of adaptations, beneficial to the living creature's survival in life. Some finches had huge beaks to crack the nuts which smaller birds couldn't manage; bloodhounds had an extraordinarily refined sense of smell, to earn their keep by hunting criminals. But how did these special abilities come about originally? How did variation in living things happen?

A further question followed. Let us accept that the world is full of individuals that vary from one another. But once these little variations have happened in an individual, could they become 'fixed', so that the advantage of the new variation could be passed on to the next generation? Darwin described his vision of this process: 'in ten thousand years the long-legged race will get the upper hand.' A new 'race' – perhaps a new species – would have been established. Darwin began to think of the multitude of variations he saw in nature and in artificial breeding, not as 'flaws' in God's creation, or divergences away from perfection, but as something incredible: the key to how species might change over time.

In July 1837, Darwin began a new notebook, bound in brown leather, with a B written on the cover. This collection of apparently random facts, scribbled on page after page of a nondescript notebook, was in fact the outward manifestation of a huge amount of systematic thought. Darwin was sorting through what he knew, retaining some of what he had learnt at university, rejecting some, and making new links between what he had seen while voyaging, and what he had learnt after he returned. He was now close to becoming a transmutationist: someone who believed that one species might slowly change in time into another.

So to his two original queries, he added one final question: if everything evolves from something else, how are all the species in the world related to one another? Darwin thought of something like a coral reef, growing from a single original point, branching and growing sideways as time went on. Here is what he drew:

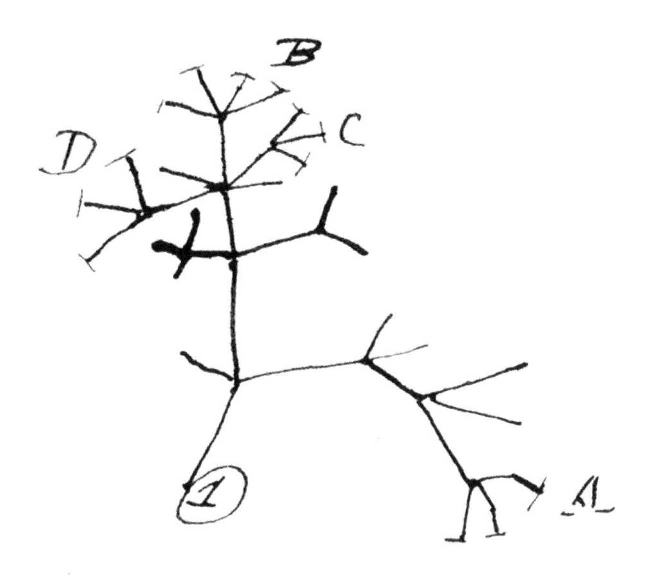

A branching network of relationships, this single tree contains Darwin's idea that all life sprung from a single ancestor, evolving and diversifying as time passed. And above it, he wrote in his distinctive scrawly handwriting, 'I think'.

Darwin wasn't just theorising. He was also energetically collecting hard evidence from many different sources. Mr Yarrell, an enthusiastic dog-lover from the Zoological Society; Hunter, an eighteenth-century expert on the relationship between dogs, wolves and jackals; Professor Thomas Bell, the man who had written the list of dog breeds in his book *British Quadrupeds.* Darwin read Anderson's *Recreations In Agriculture,* Hamilton on the dog, Colley on the care of cattle, and, by no accident, the writings of Erasmus Darwin.

In particular, he looked to practical men: breeders of prize cattle and pedigree dogs. Darwin had always believed that domesticated animals and plants – dogs, but also cats, pigeons, bees, wheat and barley – would provide an important starting point. 'At the commencement of my observations,' he wrote much later, 'it seemed to me probable that a careful study of domesticated animals and of cultivated plants would offer the best chance of making out this obscure problem.' Darwin wanted to know how variation happened, so why not consult those who seemed to have variation under their control? Certainly, variation seemed to happen most elaborately when living things were domesticated: 'The most favourable conditions for variation seem to be when organic beings are bred for many generations under domestication,' he observed.

Darwin also wanted to see species change in action, but suspected that the process in nature was too slow and too minute to be observable. So he turned to breeders, hoping to learn from their analogous process of selecting animals and plants for desirable qualities. He hoped to find someone who could show him how those useful longer legs might have come about.

Darwin was particularly industrious about finding experts in fields that interested him, and then quizzing them for relevant information. His way of working was to locate men who possessed lifelong acquaintance with a subject, and then to put his questions to those men, ferreting out their opinions and posing innocent queries which might feed his theorising. This method of research would become a feature of his work, something he stuck to all his life.

But more deeply it was also a philosophical commitment to a way of working that emphasised the collaborative, the accumulative,

SMOOTH-COATED COLLIE.

the collection of facts into a central metropolis (his study) and their subsequent assessment there. He found men who were experts on particular genuses of plants; he found men who knew all there was to know about fancy pigeon breeding; he chatted to dog men, men who had a good eye for a pup, who appreciated blood lines and who could enthuse about the shape of a fine hind leg. And then he gathered up all they knew and made use of it.

Darwin started another notebook, and then another, and then another. These notebooks became the most secret, the most precious repositories of his thinking. Many years later, while writing his autobiography, Darwin looked again at his secret notebooks. What amazed him then was how hard he had worked, an old man's respect for his younger self's energy: 'When I see the list of books of

all kinds which I read and abstracted, including whole series of journals and transactions, I am surprised at my industry.'

Again and again, dogs came up, and Darwin was particularly interested in these examples. One reason is simply that he had a good selection of dog experts to draw upon. Another is that it was a subject that interested him personally. And finally, it was a field that had its own set of intriguing problems: dogs are amongst the most varying of animals, in terms of size and shape. But it was still tantalisingly unclear to Darwin where all this variation had come from.

In particular, there was much debate about the original ancestor of the dog. There was no obvious wild ancestor for the dog, leading some to speculate that dogs were tamed wolves, jackals or coyotes. Domestication of the dog must have taken place in the prehistoric past, so there were no records to tell the story. A wild ancestor that presumably looked like a wolf had been developed by human breeding into dogs that varied from the gigantic Newfoundland to the tiny Pekinese. Perhaps the breeding of dogs held one key to the question of how that breadth of variation came about.

Darwin worked hard. He went back and found evidence about the dogs that lived amongst the ancient Egyptians, trying to take the story of dog breeding back in time some four thousand years at least. He read up on the question of the wolf and the dog, marking the pages of his books as he read. Outstanding problems became clear: if wolf and dog really are the wild and domesticated versions of the same species, why don't they cross-breed readily?

He was particularly fascinated by greyhounds, and the greyhound

THE GREYHOUND.

was the example he returned to as he tried to understand how the speed and accuracy of a hunting dog might have developed. It seemed to him a perfect test of how variation arose, and how a particularly advantageous variation, like long legs, might let an individual have more success in surviving to pass on its advantage to the next generation. Darwin wasn't only interested in dogs, but dogs were an important part of the process of him coming to complete his theory.

The more he studied, the more questions occurred to him. What is a species? Darwin asked. Is it an abstractly definable thing, which exists in reality? Or is it an artificial construct that human naturalists have seen in nature? Do species and varieties exist along the same continuum, with species merely a more defined version of the same kind of distinctiveness?

This question in particular Darwin struggled with. He took heart from the fact that experts argued that species did have real existence: 'As Gould remarked to me,' Darwin wrote, 'the "beauty of species is their exactness".' Gould had impressed Darwin with the profound sense of distinction he had observed between the various Galapagos birds. But Darwin had a further point to make, because he had always noticed that breeds of dog bred true: 'but do not known varieties do the same, may you not breed ten thousand greyhounds and will they not be greyhounds?' Darwin was beginning to confront all the most confusing questions biology had to offer.

Darwin's progress with these questions was marked in his notebooks, where he noted down details from many different sources. One of Darwin's favourite dog contacts was William Yarrell. Mr Yarrell was famous for being an excellent shot, some said the best in London. He was also an expert fisherman and wrote a standard volume on British birds, full of evocative description, especially where the colours of bird plumage was concerned. Darwin met Yarrell through the Zoological Society which held regular meetings not at the zoo itself in Regents Park, but in its museum at No. 28, Leicester Square, tucked in between panoramas, dioramas, and other popular tourist sideshows.

William Yarrell had become rich in the business of newspaper distribution. As a wealthy man, he could afford to indulge his great spare-time passions: fishing, shooting, and breeding better gun dogs. We know Darwin valued his chats with this well-informed gentleman because his early secret evolutionary notebooks contain information quoted direct from Yarrell, with his name attached.

One of the areas he quizzed Yarrell about concerned the expert breeder's experience in relation to this question of inheritance. What did Yarrell think about how offspring came to inherit their parents' attributes? Yarrell answered clearly: he thought older-established varieties had more strength in influencing the cross than newer ones.

Another of Darwin's dog correspondents was his cousin, William Fox. The two became friends when they were at Christ's College, Cambridge together. The slightly older Fox helped Darwin in his first year, introducing him to fellow students interested in natural history, and coaching him for his exams. Both followed the training to become clerics, but only Fox went on to make the Church his life, but nonetheless they remained friends their entire lives.

At Cambridge, the pair spent many happy hours hunting, for both game and beetles. They were accompanied by their dogs Sappho and Dash, who were fondly mentioned in the letters that criss-crossed between the two. Darwin also kept up a correspondence with Fox during the years after he left Cambridge, with Darwin away on the *Beagle,* whilst Fox married, and found his first job.

Darwin bemoaned the fact that his friend had not become a professional naturalist, but it was never to be. Fox was for the Church. However this didn't stop him having a deep interest in the natural world. By 1837 Fox had become Rector of Delamere, in Cheshire, and had a little more free time to pursue scientific hobbies. Animal breeding was something that had always particularly interested him; he had begun with chickens and now returned to the subject in his letters to Darwin, much to his friend's delight.

Fox was one of the sources most heavily quoted in Darwin's first transmutationist notebooks, and Darwin wrote to Fox, fondly and prophetically; 'I am delighted to hear, you are such a good man, as not to have forgotten my questions about the crossing of animals. It is my prime hobby and I really think some day, I shall be able to do something on that most intricate subject species and varieties.' Fox and Darwin always exchanged letters full of deep respect, yet Darwin's theory never won Fox round; Fox remained a believer in God's creation to the last.

In November 1837, a year after returning from his travels Darwin visited Fox on holiday in the Isle of Wight. They spent a happy few days discussing in detail what knowledge they should like to have about dog breeding, and where they could find out. As a result Fox wrote to John Howard Galton, who was a noted blood hound breeder.

Darwin wanted to know about how breeders kept their breeds 'true' – meaning how they kept the traits they wanted in their dogs, and how they prevented other traits creeping in. The main technique used by breeders to retain a desirable trait was 'breeding in' – breeding one dog with another that it was closely related to. However, Galton was quick to point out to Fox the disadvantage of this method: 'Another consequence of breeding in and in is that the animals become prematurely old'. Galton seems to have

RETRIEVER.—*Canis familiaris.*

meant that pedigree dogs are more prone to illness and early death than mongrels.

Galton was not the only bloodhound enthusiast to feature prominently in the early notebooks: there was also 'Mr Bell of Oxford Street'. Jacob Bell was a pharmacist by trade, in a substantial family firm. His father's good financial sense had built up a considerable business, leaving the son free to be a patron of the arts and attend meetings of learned societies. Bell was a particular patron of Edwin Landseer, who became famous for his animal paintings in Victorian England.

Landseer repeatedly painted Bell's bloodhounds. Bell was a keen and very particular breeder – as another Bell, Professor at King's College, wrote, 'the race has been gradually diminishing and is now very rarely to be met with in its purity. Amongst the very few instances of its presence existence, I may mention a fine breed in the possession of Mr J Bell, of Oxford Street, who retains them in great purity.' Bell paid Landseer handsomely, but also gave him much useful advice on his business affairs such as the enforcement of copyright on unauthorised engravings, a sore subject to Landseer at the time.

The first time Landseer painted one of Bell's dogs was *Sleeping Bloodhound* of 1835. Away from the business, Bell lived in a large house in Wandsworth. In 1835, his precious dog Countess fell more than twenty foot from a parapet on the house's substantial facade. Fearing that Countess was dying, Bell set off with her, not for the vets, but for Landseer's home in St John's Wood, to have him paint the dog. Fearing her imminent death, he wanted a Landseer portrait to preserve her memory. Landseer did as he was asked; fortunately, Countess survived.

Within just a few years, Bell commissioned another dog painting from Landseer: the celebrated *Dignity and Impudence* of 1839. In real life the two dogs were called Grafton, the bloodhound, and Scratch, a West Highland terrier. Despite the peaceful appearance of the scene, Grafton was prone to undignified behaviour. The last straw came when he viciously attacked another dog when they were locked in a stable together overnight; Bell threatened, next time 'I will shoot him'. In 1839 Bell also commissioned Landseer's brother Charles to paint one of his bloodhound bitches with her pups,

DIGNITY AND IMPUDENCE, SIR EDWIN LANDSEER 1839

climbing over her back as she lay in the straw.

Bell's pure blood lines were a joy to dog-lovers, but they were of use to Darwin because it meant that he had particularly good breeding records. This was of special interest to Darwin, who referred to information gained from 'Mr Bell' several times in his notebooks and letters. Bell, like Yarrell, was added to the list of trusted contributors, whose experience in breeding, and careful record-keeping, marked them out as extremely useful to Darwin. Not just dog men, either: William Tegetmeier, a pigeon-loving journalist, and George Tollet, a livestock-owning neighbour of Darwin's uncle, were also counted amongst their number.

Darwin had collected many examples, and begun to think out his theory. But now he had to flesh out how his mechanism might work. For this, he took from two thinkers in particular: the first was the geologist Charles Lyell, who gave Darwin his sense of deep time; the second was the Reverend Thomas Malthus, whose gift was a vision of the scrabbling over-fertility of the natural world.

Charles Lyell was ten years older than Darwin, and already a geologist of great repute. One volume of *Lyell's Principles of Geology* accompanied Darwin on the voyage of the *Beagle,* and it made for illuminating cabin reading. Lyell was no transmutationist himself, and yet his insights into the age of the earth eventually helped Darwin to solve the problem of how one species might become another.

Darwin had begun by thinking about very tiny changes in species which would accumulate over time. The size of the changes explained why naturalists hadn't been able to see transmutation taking place: the differences, when viewed day-to-day, were simply so small that no human being would notice them. It was only with the long action of time that there could come to be visible distinctions of the kind taxonomists used to differentiate whole species.

But in order for such tiny changes to make a creature that looked like a weasel evolve into a creature that looked more like a dog, enormous amounts of time must have been involved. Yet science had a problem here: the problem was Genesis. The book of Genesis gave the ages of all the patriarchs, and together with other information in the Old Testament, a precise length of time since the Creation could be calculated. The calculation was first done by Bishop Ussher around 1650: the world had been created just six thousand years before, on the eve of 23 October 4004 BCE. But six thousand years was definitely not enough time for the kind of changes Darwin was talking about.

Darwin's solution was to turn to Charles Lyell. At university, Darwin had been taught that only cataclysmic earthquakes and floods could have made the dramatic shapes of geological landscapes. But Lyell argued that the earth was changing on a far slower scale. Mountains were created, canyons were eroded, glaciers scoured landscapes. Yet all of this dramatic change was achieved by everyday processes such as the gentle flowing of rivers. All it required was time. The simple flow of water could erode deep canyons and eat its way through mountain ranges, changing the whole appearance of the earth's surface over thousands or even millions of years. Lyell could come to only one conclusion: the geological processes that shaped the earth's surface were incredibly powerful, but terribly slow. The earth must be many millions of years older than Bishop Ussher had imagined.

Darwin began thinking that Lyell's sense of the enormous age of the earth might also be applied to the history of life. If geological processes could act over millions of years, why couldn't species change over those sorts of timescales too? With millions of years over which transmutations might have taken place, Darwin could imagine far greater changes and how they came about. For example, Darwin had tackled small changes, as a species slowly evolved those longer legs. But what about the process of a simple one-celled animal evolving into a complex being like a dog or a human being?

Thinking in a Lyellian way, over millions rather than thousands of years, Darwin found it much easier to envisage that dogs, giant extinct mammoths, platypuses, blue whales, bats and humans all originally descended from a single ancestor, far, far back in the history of life on earth. In a million years, a lot could happen. And recent research reveals that the last single ancestor shared by every mammal probably lived around two hundred and twenty million years ago. The first placental mammal, a mammal which incubates the young by being pregnant, dates from just over a hundred million years ago. Before that, it would no doubt have amazed and delighted Darwin to know that the earliest mammals – the ancestor of both human beings and our dogs – laid eggs.

Having adopted the idea of Lyell's long timespan, Darwin's theory still lacked an engine of change – a force that would take individual variations and push them to spread throughout an entire population. In search of relaxation while he pondered the mechanism, he picked up a book that would lead to the final piece in his puzzle. It was September 1838.

Darwin was no trendsetter when he began reading the Reverend Thomas Malthus's bestseller *An Essay on the Principle of Population.* On the contrary, Darwin was trailing well behind fashionable opinion, for the book had been published in 1798 and had gone through six editions, each selling more than the last. Malthus wrote the book while working as a rural curate in Surrey; yet it became one of the most influential texts in the history of political economy. Malthus, thought hard-minded Victorians, explained why Poor Relief didn't work, and why it would be better in the long run to let the Irish starve.

Malthus's book scrutinised the wars, famines, price crashes and economic shortages which troubled mankind and coolly pointed out that whatever the variables, whatever the conditions, whatever the differences, two things would always be the same. When food supplies (which he called 'the means of subsistence') increased, they could never increase faster than arithmetically: by a certain percentage a year. But human populations, said Malthus, increase exponentially. A couple produce five children, each of those children produce five children, each of those grandchildren produce five children and suddenly one married couple have a hundred and twenty-five living descendants.

Thus, a human population can grow far quicker than the supply of food ever can. And as soon as populations are increasing faster than the food supply, said Malthus, checks will come into play. These 'checks' will include plagues, wars and most of all, starvation and famine. A new generation of children die; there is enough food to go round. For Malthus it was simply a question of the mathematical patterns.

Darwin's insight whilst reading Malthus concentrated on those checks. For Darwin, the natural world was a place of incredible competion for food, for security, for a chance to reproduce. Darwin took Malthus's idea of many more individuals being born than would ever be able to survive, each competing against the other for the basic right to continue living. He considered his own experience of the natural world: the huge masses of frogspawn in spring ponds, the large litters of farm cats. This image of a seething mass of individuals stayed with Darwin.

Lyell had argued the world was much older than previously thought, and that geological processes were much slower: from this, Darwin developed his sense of the incredible number of millenia over which species change took place. Malthus, on the other hand, was famous for his statement of how quickly species can breed, just like rabbits: it inspired Darwin's idea of the swarming fertility of nature, as it produced millions of eggs, seeds, tadpoles and puppies compared to the small numbers of surviving organisms. Both men worked an essential alchemy for Darwin, on all the material he had gathered.

Finally Darwin's theory had come together into some sort of recognisable shape. A theory of descent with modification, powered by natural selection. A wolf was born with longer legs so that it could slightly outrun the rest; it consequently found it marginally easier to catch food than other wolves. Natural selection meant that the long-legged wolf had a better chance at life than its shorter legged

WOLF.—*Canis lupus.*

rivals. It was therefore able to rear more young, more successfully. As a result, it passed its long legs and excellent running skills onto more offspring than any other wolf in its generation. Its offspring were also better at running, and better at catching food. And so the numbers of wolves with long legs increased as a proportion of the whole population. The modification was the longer legs, and natural selection did the rest.

One of the biggest problems Darwin now faced was exactly how a more advantageous characteristic, like those longer legs in a hunting dog, would be passed on to offspring. For a theory of transmutation to work, there must be a way for an animal or plant to pass on its slightly better adaptation to its descendants, which would very slightly change what the species looked like to an external observer, such as Gould who had classified Darwin's finches.

The problem was that those crafty taxonomists always managed to recognise 'a species'. In practice, species were not blurred by small marginal groups on their way off to form some new, better species elsewhere, brandishing the exit visa provided by longer legs or better eyesight. Species were real entities, identifiable in the wild, and not blurring into others through subtle gradations. They were clear, isolated communities with special, unique characteristics.

This was one of the most lasting objections to Darwin's theory. Darwin didn't know about genes; he had been educated in an environment where most theories of heredity argued that a dog's puppies were already preformed in her ovaries when she was born, with the puppies' own puppies already preformed inside them, too.

Darwin didn't believe in this 'preformation' theory, but he had a problem knowing what to replace it with. If one member of a huge population had the luck to be born with those slightly longer legs, surely the advantage would be quickly swamped when they reproduced with other members of the population? Until the science of genetics developed in the early twentieth century, followers of Darwin's theory simply had to go on a hunch that creatures did have a way of 'holding onto' advantageous adaptions, and accept that it was not yet understood.

But in the meantime, Darwin was now troubled by a more pressing, more domestic preoccupation. Five years away and he was now a grown man, with a comfortable income to live on for the rest of his life. And returning to England, he had spent more time with the Wedgwood side of his family, picking up his acquaintance in particular with his cousin Emma.

Emma and Charles were the same age, the late children of two very different brothers. There was something about Emma that drew Darwin. He would love her devotedly, often suffering anxiety when they were separated, and wrote of her much later, 'I marvel at my good fortune that she, so infinitely my superior in every single moral quality, consented to be my wife.'

They already shared a family, yet were divided by several important issues. Most importantly, Charles had shrugged off his mother's Unitarian influence at the moment of her death; whilst Emma had continued to be a strong believer, and was eventually confirmed in the Anglican Church to make sure of her respectability.

So when Darwin came to think, finally, about proposing to Emma, he was in two minds. Two columns on a sheet of paper summed up his ambivalence. Perhaps, knowing a little more about Darwin, we can now look at one of the more peculiar items on the list, and see it as the kind of compliment someone with a passion for dogs like Darwin's might pay. We might be able to see that far from being rather insulting, Darwin was paying his possible future wife a fond and mischievous tribute when he wrote that one of her possible advantages was that she would be 'better than a dog, anyhow'.

Man in his arrogance thinks himself a great work, worthy the interposition of a deity. More humble and I believe true to consider him created from animals.

In the month of June in the year 1842, Darwin put down his pencil and stopped writing. He had finished the first outline of his theory. Scribbling fast and missing out chunks of sentences, the thirty five pages are run through with additions and crossings-out. He set down what he thought rapidly, in sparky shorthand form: 'if he foresaw a canine animal would be better off, owing to the country producing more hares, if he were longer legged and keener sight – greyhound produced'. Darwin would rewrite the sketch several times in the years to come, eventually extending it to over two hundred pages. But all were statements of his belief that evolution had really happened.

Darwin had included in his outline the long timescale he inherited from Charles Lyell, and the population and resource pressure he'd found in the writings of Reverend Malthus. His theory had the details on breeding and heredity from the dog breeders, such as their pointer puppies who would by instinct point at stones; it encompassed the experience of the livestock men, who'd told him how they bred different cattle for meat than they did for tallow fat. And altogether it read as a fluent argument for descent with modification by natural selection. The document was finished, written out longhand, but now he paused. He didn't ask anybody to read it; he simply put it away in a carefully locked drawer.

There are many reasons Darwin paused. He was by nature a cautious person, most of the time. He was new to the London scientific community and wanted to earn the acceptance and respect of the establishment figures he looked up to, such as Charles Lyell. Darwin's new theory, full of colourful examples from his travels and the evidence of ordinary breeders, could have come across badly with conservative London; it might even have been seen as

DARWIN AT DOWN HOUSE

intentionally stirring up trouble. Particularly after the appearance in 1844 of a transmutationist bestseller called *Vestiges of the Natural History of Creation,* a blockbuster that even Queen Victoria read, but which the establishment ruled 'a foul book'. But finally, Darwin probably feared upsetting his wife Emma, who had written a tender letter to him around the time of their marriage confiding her anxieties about his future in the afterlife if he were to leave Christianity behind. It was not quite the moment to launch a theory that questioned the literal truth of the biblical creation.

So instead of plunging into the controversy of publication, Darwin immersed himself in settling down to enjoy his family. William had been born at the end of 1839, Annie in the spring of 1841, and for Darwin now the daily life of the household took over. Perhaps because they had decided not to stay forever, the couple began to like London, towards the end of their years there: Darwin wrote in 1840 'If one is quiet in London, there is nothing like its quietness – there is a grandeur about its smoky fogs, and the dull distant sounds of cabs and coaches: in fact you may perceive I am becoming a thorough-paced cockney and I glory in thoughts, that I shall be here for the next six months.'

But Darwin really wanted rural peace and quiet. And in 1842 the Darwins finally bought their dream house in the country. His first observation on the subject of his new home had a suitably botanical tone: 'I was pleased with the diversified appearance of the vegetation proper to a chalk district.' Adding to his pleasure at the diverse vegetation was his delight at the calm possibilities that life at Down offered. The Darwins would never move house again.

Yet it would be another sixteen years before Darwin would publish the theory he'd formulated before he moved to Down. Sixteen years during which seven of his children were born: sixteen years of them learning to smile, talk, walk and read, convalescing on the sofa in his study, and being a source of consolation and company for their unexpectedly fun-loving father. 'When you were young,' he wrote in his autobiography, 'it was my delight to play with you all, and I think with a sigh that such days can never return.' He also watched three of his children die; in one case, Annie, terribly painfully, noting down every detail in thrice-daily letters to her mother back at home, from the spa they had all hoped would save her life.

Sixteen years of hard work, too: first on corals, and then on barnacles, tiny sea creatures which proved an extraordinary underwater world. Each part of the process of looking at the miniature feelers and structures under the microscope fed Darwin's sense of certainty that the theory he kept in the drawer was correct. Nevertheless

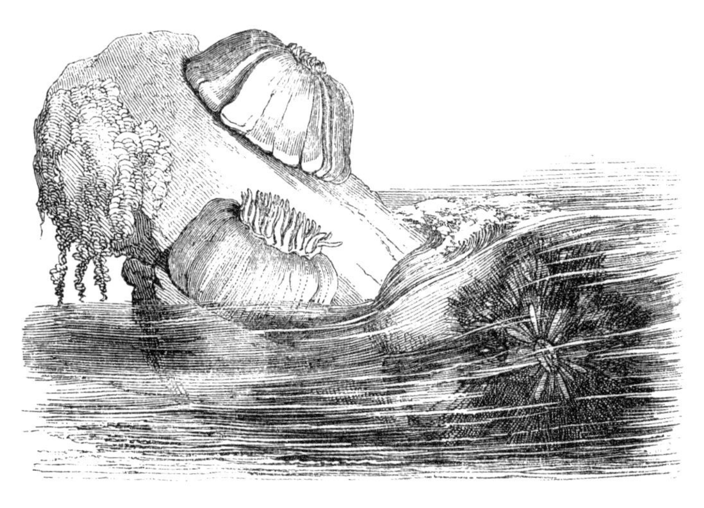

ACTINIÆ, OR SEA-ANEMONES.

he modestly wrote to his cousin Francis Galton, who had been exploring in Africa, 'the objects of my study are very small fry, and to a man accustomed to rhinoceroses and lions, would appear infinitely insignificant.'

And sixteen years of growing in confidence as a naturalist, making networks and earning the respect of colleagues. Darwin was often confined to Down by his mysterious medical symptoms: abdominal pains and nights of retching, never satisfactorily explained. But he still managed to go to London for scientific meetings, and in doing so, slowly built up a group of valued colleagues.

A very few of his fellow naturalists came to be close friends, enough that Darwin felt able for the first time to confide his thoughts on the question of transmutation. For example, in 1846, Darwin had needed someone to take over classifying the plants from the *Beagle* voyage. Someone suggested Joseph, the son of William Hooker, the Director of Kew Gardens. Joseph Hooker quickly became one of Darwin's inner circle and eventually, one of his best friends. After knowing Hooker just a few months, Darwin dared to share his heretical views; 'it is like confessing to a murder,' Darwin wrote, apprehensively. Hooker replied with an open-minded letter that convinced Darwin he had made the right choice of confidant.

But nonetheless, Darwin didn't return to working on transmutation. Through all his barnacle years, though, Darwin knew that one day he would come back to what he called his 'species work'. He just worried that after all the stress and anxiety, in the end it wouldn't come to anything. 'How awfully flat I shall feel,' he wrote to Hooker in 1854, 'if when I get my notes together on species etc etc, the whole thing explodes like an empty puff-ball'.

*   *   *

When his family were young, it was Darwin's habit to have his breakfast on his own before eight o'clock, and then go to work straightaway. He liked to grasp the freshness of that early morning moment before the day had begun to bustle. He'd take a break at half past nine, and come into the drawing room for his letters and a cup of tea, then return to work at half past ten, until midday or so. Then he'd take a walk up the garden, whether it was fine weather or not, taking any dogs in the house with him, doing the circuit five times before lunch.

A gentle daily rhythm evolved, giving Darwin the structure he needed to research and publish. The years rolled by; eight summers and winters, till he got to the point where his subject matter made him want to scream: 'I hate a barnacle as no man ever did before, not even a sailor in a slow-sailing ship.' Nevertheless, the barnacle book, when it finally appeared, confirmed Darwin's reputation as a biologist of skill and detail. He was awarded the Royal Society's Gold Medal for the work whose minute findings impressed the world of the naturalists.

And when Darwin finally returned to the species notes, he also returned to the subject of dogs. With barnacles finally finished in 1854, Darwin could launch himself back into working on the species ideas with much greater enthusiasm and confidence. He planned to write an authoritative book on the subject. But Darwin, as always thoroughly systematic, couldn't simply just publish from his existing notes. He needed to revisit his theory, checking and rechecking every possible line of criticism.

He sorted back through old notes – the notes from Mr Bell of Oxford Street; the thoughts on bloodhounds from Galton. He found new correspondents, like Edward Blyth, the curator of the Museum of the Asiatic Society of Bengal; and Hugh Falconer, the former head of Calcutta Botanic Garden. Blyth sent him details of pariah dogs, the shunned wild mongrels that hang around living off the detritus of Indian villages; Falconer wrote to him with information about Tibetan mastiffs. And still in touch with William Yarrell, Darwin said yes to one of Yarrell's suggestions: the keeping of fancy pigeons.

The interest in dogs ran alongside a new set of obsessions. Darwin wanted to study the movement of seeds on sea currents, so the mantelpiece in his study was covered in tiny germinating plants. And he wanted to examine the anatomy of unusual breeds, so he

HUNTING-DOG.—*Lycáon venáticus.*

invested in Yarrell's fancy pigeons. He wrote to his son William, who was away at school, on 25 April that the pigeon house was nearly installed, and that each pair were going to cost an eye-watering 20 shillings.

The pigeons toed an uneasy line between being experimental specimens and pets. Darwin found himself desperately wanting their skeletons so that he could study them, yet he was almost incapable of actually killing them. In the end, his cousin William Fox did much of the pigeon killing: 'you best and kindest of murderers,' he called Fox, in one of a series of thank you letters. 'Upon my word I can hardly believe that anyone would be so good-natured as to take such trouble and do such a very disagreeable thing as kill babies; and I am very sure I do not know one soul who except yourself would do so.' (Darwin had less sympathy with cats, though: in January 1841 he wrote to Fox, 'don't forget, if your half-bred African cat should die, that I should be very much obliged, for its carcass sent up in a little hamper for skeleton.')

The house was becoming an experimental station. It was also, more than ever, a clearing house for information coming from all over the globe. Long letters came from India penned by Edward Blyth, who sent back pages and pages of information about domesticated animals in India, giving details about rabbits, doves and jackals. And in correspondence with Hooker, Darwin was asking the difficult questions that would need to be posed if the book really were to be published.

In particular neither man was satisfied that they had yet understood the nature of the dividing line between species and varieties. There was genuine difficulty here (many disagreements exist on the subject even today). It's now understood that species and varieties are relatively fluid categories, and that in the end, the distinction can be to some extent a matter of an individual scientist's temperament.

Darwin and Hooker quickly recognised that some naturalists by nature were what they called 'lumpers', who would tend to define species as broadly as possible, by including a great range of variation. The opposite of a lumper was a 'hair-splitter', who would tend to place individuals with very variant characteristics into a completely new different species. Hooker was a self-diagnosed member of the first camp, exasperated by botanists in obscure colonial outposts sending 'new species' to Kew that Hooker saw clearly occurring elsewhere. He once wrote wrily of his colleague George Bentham, 'Well, he has turned out as great a lumper as I am! And worse'.

Despite the difficult questions that lacked answers, slowly but surely Darwin was putting together his 'big book'; a full account of the theory he had come up with all those years before, but ready now for publication. Still he prodded away, never quite convinced he had enough information on the crossing of different dog breeds, or the kinds of plants that grew on remote oceanic islands.

Yet the peaceful, scholarly publication he imagined was never to be. The first noises of trouble came in 1856. Sir Charles Lyell became very concerned after reading a paper in the *Annals and Magazine of Natural History*. The *Annals* was a journal for serious naturalists, edited by Joseph Hooker's distinguished father William, amongst others, providing a place to read articles ranging across all natural history, as well as summaries of the year's proceedings in the British and Irish scientific societies.

Lyell had been alarmed by a piece written by Alfred Russel Wallace, a young collector based in the Malay archipelago, suggesting that 'every species has come into existence'. On detailed examination, it seemed to be dangerously close to what Darwin had told Lyell in private about his own theories. Lyell urged Darwin to hurry up and publish before someone else got there first. It was May 1856. Darwin kept on working at his own speed. He ignored Lyell's warning.

As a result, on 18 June 1858 Darwin received the news that every scientist dreads: Alfred Russel Wallace had got there before him. For a year or so Darwin had actually been exchanging letters with

Wallace, who had become one of Darwin's many local experts, seeking out skins of 'curious' poultry and sending them back in cumbersome packages to Down. Darwin and Wallace wrote letters that took months to travel between Britain and the Far East, but the two men found that despite the distance, they 'thought much alike' about the great issues of the time. Wallace was delighted finally to find a fellow speculator to share ideas with, especially one of Darwin's prestige.

Wallace began developing a longer version of his own theories, intending to send it to Darwin, written out in longhand. At the same time, though, Darwin continued with his own work on his species book not realising that Wallace was working on something dangerously similar, thousands of miles away. In early 1858, Darwin had accumulated two hundred and fifty thousand words of his 'big book' – fourteen chapters, laid out in piles in his study. But in June he received the letter from Wallace. It spelt out a catastrophe. Darwin's Asian collector had trumped him: the four-thousand-word letter outlined an entire parallel theory of evolution.

Darwin wrote that day to Lyell, who had been all too correct when he had warned Darwin to publish before somebody else did. Wallace had not just got there first; Darwin wrote that Wallace had even stated his ideas in the same terms: 'if Wallace had my manuscript sketch written out in 1842 he could not have made a better short abstract!'

Darwin was in agony. He didn't think he could now justify publishing any of his own work, because it would mean stealing priority from the young man faraway in the East. 'I cannot tell whether to

ALFRED RUSSEL WALLACE

publish now would not be base and paltry… I would far rather burn my whole book than that he or any man should think that I had behaved in a paltry spirit.' It was Lyell, the old friend, who found the way forward. Both papers should be published jointly. A Linnean Society meeting was scheduled for 1 July 1858.

But for the crucial meeting, Darwin stayed at home. In the evening of 28 June his and Emma's youngest child, the baby Charles, had died of scarlet fever. The fact that the papers had been read, to little initial interest, hardly seemed to register. The family must have been miserable, and desperately anxious about all the other children too. Preoccupied with domestic matters, Darwin could hardly think straight. A holiday in the Isle of Wight for recuperation was organised. On his return, though, Darwin got straight to work on a short book for general publication, summarising all his work on the species question. The book he now began writing was the one which would make him famous: *On the Origin of Species by Means of Natural Selection.*

Darwin's intention in the new book was to explain his mechanism of 'natural selection.' The book would deal with how variation came about, the struggle for existence between varying individuals, and the question of how diverging varieties might come to form new species. It also tackled many possible objections to the theory, such as the lack of fossil evidence and the evolution of complex organs like the eye. But, true to the way in which he had come to his theory, Darwin chose to begin with a chapter about variations in domesticated animals and plants, the subject he'd been researching for years. Chapter one's very first sentence turns straight to the question of why we see far more variability in domesticated species than we do in those observed in nature.

Think of dogs, said Darwin. Even today, experts argue, dogs have amongst the greatest range of variation in size and shape of any vertebrate species that has ever lived. Darwin tried to explain why. Under domestication, Darwin thought, the whole organism became what he called 'plastic'. Domesticating animals, Darwin believed, was a process that allowed a whole range of variability to be revealed and to be brought out into the open, which is hidden in the wild ancestors: variability that selection could act upon.

Where did all this come from, asked Darwin? He thought that the domestic dog was probably descended from a range of wild canid ancestors, each contributing something to the mix. But whatever the truth of canine origins, the crucial question for breeders was how to get more of the desirable qualities they wanted in a particular animal. For Darwin, the fascination was how that process of picking and choosing worked. 'When we compare the many breeds of dogs, each good for man in very different ways,' said Darwin, 'we cannot suppose that all the breeds were suddenly produced as perfect and as useful as we now see them.'

The key, Darwin argued now, was 'man's power of accumulative selection: nature gives successive variations; man adds them up in certain directions useful to him.' This was exactly the process that Darwin had observed, time and again, since his early childhood. Darwin wanted his readers to settle down and accept the wisdom of the practical men, the breeders whose work he was discussing, because he was next going to make an analogy between the work of the breeders, which he called 'artificial selection', and the enormous creative power of the force he called 'natural selection'. For Darwin there was just one significant difference between selection as performed in the farmyard by a cattle breeder, and selection by nature itself:

> Natural selection ... is a power incessantly ready for action, and is as immeasurably superior to man's feeble efforts, as the works of nature are to those of art.

God's perfectly formed Creation has disappeared from this account; in Darwin's vision, nature is simply one huge and incredibly skilled breeder.

Darwin had put together all the necessary parts of his theory. Plants and animals vary. Many more are born than will survive to reproduce. Each is very slightly different from the next. In such competitive circumstances, the tiniest advantage is enough to ensure one individual survives or that one has slightly more offspring. As the advantaged individual reproduces, it passes on to slightly more descendants its lucky inheritance. The lucky inheritance spreads slowly through the population. Over the huge span of geological time, a tiny primitive blob of an organism could become something more sophisticated; develop eyes and begin to see; develop limbs and begin to swim; develop lungs and move out into the open air. Little by little, bit by bit, Darwin argued, species evolve.

These days we'd expect a scientific book, even one written for the general public, to be backed up with statistics, diagrams and tables. One of the first things you'll notice when you open a copy of *The Origin of Species* is that there are none of those things. Our first guess might be that the book was intended for a very popular audience: but *The Origin of Species* was written for a highly educated gentlemanly elite, not the paperback-buying public of today. So what explains the way in which Darwin presented his crucial argument?

Here is an example. In the chapter on variation under domestication, Darwin discusses all kinds of domestic animals and their wild relatives. Darwin had become particularly interested in the stripes of different members of the horse genus, Equus, such as quaggas, Indian kattywars and zebras, as well as ordinary donkeys. He began looking for the stripe in ordinary horses, too, hoping to prove a family resemblance:

> With respect to the horse, I have collected cases in England of the spinal stripe in horses of the most distinct breeds, and of all colours ... My son made a careful examination and sketch for me of a dun Belgian cart-horse with a double stripe on each shoulder and with leg-stripes; and

> a man, who I can implicitly trust, has examined for me a small dun Welsh pony with three short parallel stripes on each shoulder.

When a scientist puts together an argument like this, they are seeking to convince their readers, just like any other speaker. Darwin wanted to convince his friends at the Linnean Society, but he was also aiming to win the kind of intelligent general reader who had gone out and bought that 'foul book', the *Vestiges*. But instead of using tables and statistics, Darwin put together example after example.

You can think of *The Origin of Species* as being a bit like one of those Victorian houses with hundreds of knick-knacks on every available surface. The initial impression of many people reading this book for the first time is of profuse over-ornamentation; exactly how you can feel in one of those stuffy Victorian interiors.

Yet if we view this as a disadvantage, perhaps as something stopping us getting down to the nitty-gritty of Darwin's theory, we are missing the point that all these examples actually make up the fabric of Darwin's argument. Darwin's sources were global, drawing in evidence from all round the world. He went searching for spinal stripes on horses of all kinds. This cart-horse was Belgian: these little details underline the book's claim to be international in its reach.

Darwin also suggests that he has amassed examples not just from across the earth's surface, but also over a considerable length of time. He emphasises his acquisitive search for facts: for horses, he says, 'I have collected cases of the spinal stripe in horses of the most distinct breeds, and of all colours', implying that there is probably an equivalent parallel collection of cases for each paragraph in the book.

And finally he makes sure to emphasise the credibility of his witnesses. Darwin had a huge network of correspondents. But it was important for Darwin to convince the reader that his correspondents were trustworthy men themselves. 'A man, who I can implicitly trust', tells us that Darwin's authority rests on a solid bedrock.

In all three ways, which all occur in this single paragraph, Darwin is trying to say: 'trust me, I know what I'm talking about'. These days, we'd expect a scientist asking for our trust to be armed with tables, charts, statistics and diagrams. Darwin did as much research, perhaps more, as any modern-day scientist would before publishing a ground-breaking paper, but even today, general readers of science would have to trust that a scientist had done their work properly. And here is Darwin, asking for equivalent confidence from the reader in his introduction:

> I cannot here give references and authorities for my several statements; and I must trust to the reader reposing some confidence in my accuracy. No doubt errors will have crept in, though I hope I have always been cautious in trusting to good authorities alone.

Yet Darwin was not quite a scientist; historians often use the word 'naturalist' to describe Darwin, to differentiate him from later nineteenth-century professionals who held paid academic positions in universities and whose books did include statistics and tables. Once we know this, the addition of example after example makes sense. Darwin was hammering home his point in fact-filled prose,

ZEBRA.—*Asinus Zebra.*

the best way a Victorian amateur knew how.

There is one more thing to notice about the way that Darwin wrote his book. He begins with the section about artificial selection, as already noted. This is because he wants to use the idea of the breeder as an analogy, so that we can understand later in the book what he means when he uses the term 'natural selection'. But the other thing that it does is familiarises the concept of selection: it brings it into the home, onto the hearth rug and curls it up in front of the fire.

So the dogs, pigeons and cattle that appear at the beginning of *The Origin of Species* are not there by accident. The dogs are there as evidence of the long experience of breeders, upon which Darwin has drawn. The dogs are there to allow readers to follow the arguments by relating them to a subject with which they'd already be familiar. And finally the dogs are there to make this new theory seem comfortable, a gentle homely theory, not the stuff of a godless atheism that threatened all that Britain held dear.

The Origin of Species is often cited as a book that is difficult to read. There will always be moments when tackling the *Origin* which remind the reader of Huxley's description of the book as 'a mass of facts crushed and pounded into shape.' But it's also full of beautiful imagery. When Darwin discusses natural selection, he asks the reader to focus on the example of a single wolf.

This wolf, says Darwin, hunts successfully, sometimes by being craftier than its prey, sometimes by being stronger, and sometimes by being faster. Darwin chooses to focus in on 'that season of the year when the wolf is hardest pressed for food.' We are there with the wolf, in the cold, waiting and hungry. 'I can under such circumstances see no reason to doubt,' wrote Darwin, 'that the swiftest and slimmest wolves would have the best chance of surviving ... I can see no more reason to doubt this, than that man can improve the fleetness of his greyhounds by careful and methodical selection.' The parallel, said Darwin, holds.

> Natural selection is daily and hourly scrutinising, through the world, every variation, even the slightest; rejecting that which is bad, preserving and adding up all that is good; silently and insensibly working, whenever and wherever opportunity offers.

And so, just as a human breeder constantly breeds his hunting dogs to have sharper eyesight and longer legs, so does nature.

Darwin's illness during the period he was writing the book has sometimes been put down to anxiety about the religious aspect of what he was writing. Whether or not this is true, the theological implications of the book were central to its reception. One vital task Darwin had to accomplish in his book was to tiptoe between the needs of his theory, his argument, and the strongly-held religious beliefs of some members of his audience. At several points in the *Origin* Darwin tackled head-on the question of what his theory meant for those who believed in a strictly biblical view of creation.

His technique was to be very gentle but persistent; deliberately juxtaposing his own view with the interpretation that a reader with a belief in creation would take. For example, return to that discussion of the stripes that occur on different species from the horse genus Equus, such as donkeys and zebras. Someone who believes in creation, says Darwin, would have to believe that each Equus species had been created with a tendency to vary in the same manner, each producing stripes in certain situations. This means believing that God has created the world specifically to contain hints of family relatedness between species in it, even though those hints are fraudulent.

For Darwin this belief could only be a wrong turn, even if one believed in God. 'To admit this view is, as it seems to me, to reject a real for an unreal, or at least for an unknown cause. It makes the works of God a mere mockery and deception.' Darwin argues quite clearly that to maintain a belief in the creation of species, in the face of all the evidence presented in the book, was actually to do a disservice to God by suggesting that God would try to deceive the human race. The works of God would not be deceitful, Darwin suggests.

So what should religious people think about the similarities they saw in nature? Even the most firmly religious naturalists such as Richard Owen, admitted that the species they anatomised resembled others. What explained this closeness of appearance? Owen argued it was the expression of the Creator's plan, dividing all species into related groups whose similarities were there to remind observers of the harmony in God's creation. Darwin, on the other hand, was clear. All of life had sprung from a single point in the past. The resemblances were not due to divine intention; they were due to family likeness. Members of the horse family had stripes because they were actually a family. And their shared anatomy made it clear: somewhere in the deep past, sea lion and dog were related.

Yet the objections to the theory of descent with modification were not just religious. Darwin dealt head-on in the book with the possible scientific objections to his theory. 'Some of them are so grave', he wrote, with great candour, 'that to this day I can never reflect on them without being staggered.' Some would later cite this of evidence of the weakness in Darwin's work; to others, it is a sign of his inherent modesty and willingness always to learn more.

His list of objections was simple. Firstly, where are the transitional forms, he asked, between one species and another? Why, if evolution occurs as a slow process, do we see each species so clearly, without 'blurring' between one and the next? And how could natural selection gradually produce a highly complex structure like an eye?

BLACK WOLF.—*Canis occidentālis.*

Almost two-thirds of the book is caught up in addressing these objections, winding around the world to form a global account of the process of evolution, from fossil beds in the sub-Himalayas to the flora of New Zealand. Once more he is trying to convince the reader that he has taken every kind of evidence into account: across time, in the form of fossils, and across space, by using the networks of the British Empire and of the Royal Botanic Gardens to marshal specimens and facts from all across the world.

It is this restless ranging across the globe, flicking from the island chains of Indonesia to Panama in three sentences, which gives Darwin's work its final and most convincing note of authenticity. We travel the world with him, taking giant-sized footsteps which

BULL-DOG.—*Canis familiaris.*

can span a whole continent at once: 'We ascend the lofty peaks of the Cordillera and we find an alpine species of bizcacha; we look to the waters, and we do not find the beaver or musk-rat, but the coypu and capybara, rodents of the American type.'

Critics came up with many different objections to the *Origin.* Some critics of Darwin's theory were particularly suspicious of claims that natural selection's gradual, incremental shifts would be able to produce complicated structures: their prize counter-example was the eye. How could such a complex structure, with lens, cornea and retina have evolved slowly from a single light-sensitive cell?

These critics scoffed at the idea that a light-sensitive cell might have been the beginnings. What function would a retina have had before it was a complete retina, and how could it have worked before it was complete? If it couldn't have benefited the animal in intermediate stages, there would have been no competitive advantage for natural selection to work on. Back to the drawing board for Darwin's mechanism, said these critics. And even if you believe that such mechanisms might have slowly evolved, said the critics, where are the intermediate examples in nature? Where are the birds with only half a wing, the creatures which can half-see?

Darwin answered here by pointing to the enormous, unimaginable expanses of geological time about which Lyell had been so clear. For Darwin, his nineteenth-century moment was a mere pin prick on a giant timeline that stretched for hundreds of miles in either direction. Looking for half-wings or other intermediate characters and being surprised not to find them was a bit like picking up a book, finding that the one sentence you opened the book at did not contain the letter 'x', but then proclaiming as a result that the letter 'x' did not exist.

And finally, Darwin returned to dogs once more to counter the critics. He could not show his opponents the entire history of evolution in the rocks. Neither could he point to examples of evolution in progress in nature. But dog breeders could not even show the whole history of a development of a modern breed. 'Opponents will say, show me them [intermediate forms]. I will answer yes, if you will show me every step between bulldog and greyhound.'

Darwin worked hard to satisfy his critics on the crucial questions his book raised. But for the *Origin's* readers, the most intriguing and controversial question of all was the one that Darwin singularly avoided addressing: what did all this mean about the origin of human beings?

As Mr Leslie Stephen observes, 'A dog frames a general concept of cats or sheep, and knows the corresponding words as well as a philosopher.'

With the *Origin,* Darwin changed many people's view of the natural world forever. Darwin's natural world was still a splendid one, but it was a harsh and violent one nonetheless. Published late in November 1859, the first edition appeared bound in green cloth, costing fourteen shillings, gold-blocked with the name of the publisher, John Murray. Twelve hundred and fifty books were printed: sufficient, Murray thought, to meet the demand. But the edition was already oversubscribed by the time of publication day. Murray, an acute businessman, immediately acted. After Christmas, in the freezing cold first week of January, a second edition of three thousand appeared (allowing Darwin to make a few swift corrections). A translation into German was planned. *The Origin of Species* was beginning to cause a stir.

London reviewers were tetchy, and they were quick to make leaps that Darwin had studiously avoided. Man 'was born yesterday – he will perish tomorrow,' wrote the *Athenaeum.* All the critics were preoccupied with the question of what the book meant for mankind. And Adam Sedgwick, Darwin's old tutor from Cambridge, wrote a letter in which he admitted his admiration for parts of the book, and his grief over others. If Darwin's theory were true, Sedgwick wrote, 'humanity, in my mind, would suffer a damage that might brutalize it, and sink the human race.' He signed the letter 'a son of a monkey.'

In his few moments of distraction, Darwin retreated to the world of plants, sundews and orchids, a delicate, refreshing palate-cleanser after his book on the big 'species question'. But whilst Darwin buried himself in the mechanisms of pollination in obscure members of the Vanda orchid tribe, the newspaper and magazine proprietors of Victorian London were beginning to realise that they could make good capital out of the *Origin.* But to make the best press splash, they needed to go straight for the biggest story. At just that moment the explorer Paul du Chaillu was wowing London audiences with his tales of gorilla hunting in darkest Africa. It was clear to the magazine writers which way the story must go: man had lost his dignity, he was nothing now but a monkey.

Humour magazines began to print mocking cartoons of Darwin, bent over like a chimpanzee, with an ape's body and tail. The *Punch* staff composed poetry on the subject of the gorillas. All London saw, all London laughed. The subject of the *Origin,* of Darwin, and of monkeys, became completely confused in the mind of the public. Now, it was a storm. At every step, Darwin was haunted by the allegation – the one that he had been careful never to make – that his book claimed human beings were simply animals; not higher beings at all.

Darwin tried to ignore the row, even though he rejoiced when he found converts to his theory of natural selection. He hid out at Down, walking with Tartar, the terrier left behind by the departing vicar and adopted by the Darwins. He made no public statement on the subject, leaving hectoring the unbelievers to his loyal lieutenant, the young Professor at the Royal School of Mines, Thomas Huxley. Months and then years passed by as Darwin concentrated on his plant experiments, and also worked compiling another book to include all the background research he had done in the course of writing the *Origin.* When *The Variation of Plants and Animals Under Domestication* came out in 1868, its very first chapter was 'Domestic Dogs and Cats'.

Darwin encountered significant problems boiling twenty years' worth of material down into a single text, and enlisted his old friend Charles Lyell. 'You gave me excellent advice about the foot-notes in my dog chapter', Darwin wrote afterwards in a rueful letter of thanks to his friend, 'but their alteration gave me infinite trouble, and I often wished all the dogs, and I fear sometimes you yourself, in the nether regions.'

Throughout these years, Darwin tried to avoid the human question, but the public clamour never went away. And Darwin himself became angry when educated people rejected the evolution of human beings from ape-like ancestors. 'For my own part,' he wrote, 'I would as soon be descended from that heroic little monkey, who braved his dreaded enemy in order to save the life of his keeper... as from a savage who delights to torture his enemies, offer up bloody sacrifices, practices infanticide without remorse, treats his wives like slaves, knows no decency, and is haunted by the grossest superstitions.'

The issue of slavery as America plunged into civil war in the 1860s twisted the issue further. For Darwin, the question of slavery demonstrated the warped nature of much thinking about the human race. And so it was that after finishing his book on domesticated animals and plants, Darwin began to write, finally, about human beings. During the last years of the 1860s, he began to work on *The Descent of Man*. And despite the title, *The Descent of Man* would turn out to be a book with a great deal to say about dogs.

He was now sixty years old, and worked slowly as always, picking out examples, writing more letters, checking and re-checking facts. The children were growing up now; as he wrote, he posted the chapters to Henrietta to read, on holiday in Cannes. The new book would tackle humanity, setting man firmly in the context of his other animal relatives. But it would be a different kind of book; the *Origin* had been a book of twenty years thought, a cautiously phrased argument that ended with a meditation on the beauty of creation. The *Descent* tackled bigger themes: the question of mind, the issue of morality, and even sexual attraction.

Humans as mere animals who wore clothes: it was this idea that most troubled Victorian readers of evolutionary writings. The notion wasn't just worrying for the very godly, the bishops, clerics and congregations of nineteenth-century England; it also caused concern to many devout naturalists and scientists. Charles Lyell, for example, one of Darwin's closest intellectual allies, found his friend's theories intriguing, but was nonetheless unhappy about extending the theory of evolution by natural selection to the case of human beings.

The stumbling block was the uniqueness of humankind. Objections concerned one point in particular: the human soul, which Christianity taught would be immortal after death. The soul became a key point of debate, for the gift of a soul, which symbolised the special relationship of human beings to God, presented the devout evolutionist with a huge problem: how to explain its presence in human beings, and its absence from all other creatures?

Lyell thought there must be two separate parts to a human being. There was the animal body, which he could accept had evolved; but then also a separate moral and intellectual part, which had been created by God. Such a distinction would restore the huge and uncrossable divide between the natural evolved world, and the created world of humankind. Lyell theorised that there had been a second creation moment at the birth of the human species, when God imbued human beings with their souls. Without this hypothetical addition, Lyell thought, human beings would be stripped of their unique relationship with God, and be left mere animals scrabbling in the dirt. And he was by no means alone.

Darwin wasn't particularly sympathetic to Lyell's feelings; 'I am sorry to say that I have no "consolatory view" on the dignity of man,' he wrote. Darwin was no longer as cautious as he had been in the past about expressing such views. By the 1870s the cultural landscape had changed substantially from the anxious years that he had spent desperately wrestling with whether to publish his theory; it was now time to speak out on the issue of human descent.

The Descent of Man was to be a serious book, and it was the notion of that uncrossable barrier between man and animal that Darwin now set out to tackle. His technique once more was to amass hundreds of little pieces of evidence to tackle the champions of human uniqueness. Only human beings, they said, had the ability to love, to act selflessly, the ability to hope and plan. Only humans could feel complex emotions, think abstractly or imagine. For Darwin, all these 'unique' human characters could be observed in animals too. 'My object,' wrote Darwin, 'is to show that there is no fundamental difference between man and the higher mammals in their mental faculties.' And he began collecting the facts to prove it.

But still the image of man as monkey distressed and inflamed the debates, and as it did so, Darwin vomited, tingled, ached and shivered. There was something about that particular higher mammal which seemed to make things particularly fraught. So when Darwin finally sat down to write his book about human beings, he drew a great deal of examples from other species too. And one of the creatures he used most of all when drawing comparisons between human behaviour and that of the higher members of the animal kingdom, was the creature who had been beside him all along: the dog.

Darwin set off as he meant to go on: the first chapter was entitled bluntly 'The Evidence of the Descent of Man from Some Lower Form.' Darwin aimed to convince the reader that human beings are animals, as absolutely as any other creature on earth, and part one of the book is all about putting Darwin's idea to the test. For example, how similar is a human skeleton to that of other mammals, such as dogs? Can human beings and other mammals catch any of the same illnesses? And can people, like dogs, ever wiggle their ears?

Ear wiggling, as it turns out, is an important piece of evidence. A dog can prick up its ears to allow it to hear better, but also to signal to others that it is paying attention. For most human beings, this ability is no longer necessary. Possibly, says Darwin, because we have such ease of movement of our heads. We now put our heads on one side to show we are really listening.

Yet a small minority of people still possess the ability to wiggle their ears. If human beings were separately and uniquely created, and set apart by God from all the other animals, asked Darwin, why did God bother installing this useless feature in the human ear? Surely the obvious explanation, Darwin continued, is that if humans and dogs can both move their ears, this is a good piece of evidence that they are both descended from a common ancestor which could wiggle its ears too.

Darwin continued to look at the ear: he was interested in the tiny point which some people have on the outer cartilage of their ears. Studying embryos, foetuses and apes, he came to the conclusion that this is the vestige of the ear tip, a tiny leftover from the days when we needed to be able to prick up our ears. In the folds and curves of a human ear, the traces of an animal past.

However Darwin's main priority was not looking at the anatomical similarities between humans and other animals. He was looking to list those qualities which people agreed were possessed by human beings alone: and then he was aiming to find evidence for those qualities in the 'lower animals'; animals such as dogs.

He began with happiness. 'Happiness is never better exhibited than by young animals, such as puppies, kittens, lambs etc, when playing together, like our own children,' he wrote. By starting with puppies, lambs and babies, Darwin was aiming to involve his readers in a kind of thought-experiment. In order to bring the truth of human origins home to them as effectively as possible, he was asking his readers to use their own experience and to reason from their own experience of the natural world, so that the conclusions they reached were based not on Darwin's arguments and facts and experience but upon their own.

He used his readers' experience of dogs, domestic creatures, familiar to all, leaving the inflammatory taint of monkeys far behind.

HENRIETTA DARWIN WITH POLLY

He touched on ants and bees, familiar from every moralising Victorian children's book; beavers, another favoured symbol of animal industry; and horses, a creature upon whom Victorian life depended. His gentle writing style invited the reader in, and then asked them to remember something they already knew. 'Every one knows,' many of his sentences begin, asking the reader for acquiescence, for agreement, in friendly village fashion.

Darwin began with happiness, but he quickly moved on to grumpiness. 'Some dogs and horses are ill-tempered,' he wrote, 'and these qualities are certainly inherited. Every one knows how liable animals are to furious rage, and how plainly they show it.' For Darwin, this anger was evidence that humans and their close mammal relatives are alike in feeling strong emotions in a similar way.

He considered the kindness of animals, too, as he saw it; 'Many animals,' wrote Darwin, 'certainly sympathise with each other's distress and danger... I have myself a dog, who never passed a cat who lay sick in a basket, and who was a great friend of his, without giving her a few licks with his tongue, the surest sign of kind feeling in a dog.' Kindness, argued Darwin, showed a complexity in a dog's emotional world that differed from that of human beings only by degree.

One of the most poignant moments in the book concerns the loyalty of dogs, a subject that fascinated the sentimental Victorians. 'The love of a dog for his master is notorious,' wrote Darwin; even 'in the agony of death a dog has been known to caress his master'. He quoted from the debate over experiments on live animals, a practice he had hated during his short period of medical training at Edinburgh: 'Everyone has heard of the dog suffering under vivisection, who licked the hand of the operator; this man, unless the operation was fully justified by an increase of our knowledge, or unless he had a heart of stone, must have felt remorse to the last hour of his life,' he wrote in the *Descent*.

Vivisection was a deeply controversial subject. The word vivisection meant any procedure done on a live animal, but in Darwin's time it was common for experiments to be undertaken on animals without anaesthetic. Animals had to be strapped down to stop them biting and moving: it is little wonder that Darwin found the process

TERRIER.—*Canis familiaris.*

sickening. He was by no means alone. Queen Victoria herself was opposed to the practice, and it was her support and patronage that saw the Society for the Prevention of Cruelty to Animals, established in 1824, becoming Royal in 1840.

Darwin found vivisection repulsive, but he also felt that physiology required experiments on animals. Darwin wrote to Ray Lankester in March 1871: 'You ask about my opinion on vivisection. I quite agree that it is justifiable for real investigations on physiology; but not for mere damnable and detestable curiosity. It is a subject which makes me sick with horror, so I will not say another word about it, else I shall not sleep tonight.'

So despite his sickness and horror, Darwin supported the passing of the Cruelty to Animals Act 1876, and was summoned to the select committee on the subject; cruelty to animals, he said at the Royal Commission, 'deserves detestation and abhorrence.' He agreed with the Act's moderate provisions, which set out that animals used in experiments should be anesthetised, and should only be used once before being put down. Darwin believed that medical progress was impossible without using some live animals; but he always strongly advocated the best possible treatment for the animals concerned. 'I have all my life been a strong advocate for humanity to animals, and have done what I could in my writings to enforce this duty.' Amongst all those writings, *The Descent of Man* was his most powerful testimony for treating human beings and animals more similarly, for viewing the human and the animal as worlds not separated by a gulf, but instead as a continuity.

Sometimes, Darwin's examples showed the sheer delight he found in his dogs. 'Dogs show what may be fairly called a sense of humour, as distinct from mere play; if a bit of stick or other such object be thrown to one, he will often carry it away for a short distance; and then squatting down with it on the ground close before him, will wait until his master comes quite close to take it away. The dog will then seize it and rush away in triumph, repeating the same manoeuvre, and evidently enjoying the practical joke.'

Some critics would dismiss Darwin's claim to be able to know anything about what is going on inside the head of a dog that rushes to and fro like this with a stick. But Darwin had no such reservations. For him, it was clear that the dog was enjoying the joke, actually teasing its owner. And as readers, we feel the prick of recognition. Example after example is there to convince us that if we have observed these things in the world around us, we should accept that human beings and animals have common origins.

Darwin's idea that the dog is showing a sense of humour could be dismissed as anthropomorphism, the attributing of human qualities to beings who are not human. We cannot know what the dog is doing, these critics might say, because the dogs cannot tell us. Yet for Darwin it was possible to observe dogs interacting with human speech, understanding and responding to it, as a mute person might. Darwin didn't think that the lack of speech was a serious obstacle to considering the essential community of animals and humans. 'A community of descent,' was the phrase he used in the *Origin*. Animals and human beings: one community, and somewhere, one single ancestor.

The inability to speak certainly never stood as an obstacle to Darwin's relationship with his dogs either. Darwin wrote about the emotions of animals as a scientific observer, but it is clear that he always related to his own domestic animals with tenderness and delight. The Darwins never took their dogs on holiday, and so Francis Darwin could write of his father many years later:

> He rejoiced at his return home after his holidays, and greatly enjoyed the welcome he got from his dog Polly, who would get wild with excitement, panting, squeaking, rushing round the room, and jumping on and off the chairs; and he used to stoop down, pressing her face to his, letting her lick him, and speaking to her with a peculiar, tender, caressing voice.

And whilst dogs were always Darwin's favourites, Henrietta Darwin wrote of her father's tolerance towards her own pets:

> He cared for all our pursuits and interests, and lived our lives with us in a way that very few fathers do… He had no special taste for cats, but yet he knew and remembered the individualities of my many cats, and would talk about the habits and characters of the more remarkable ones years after they had died.

Darwin saw each animal as having its own separate being, its own 'individuality'. His celebration of Henrietta's remarkable cats suggests, just delicately, that if we had asked Darwin if cats had souls, he might have answered: as much as any other creature does.

Darwin now turned to the question of imagination. Surely, this was one human talent that couldn't be said to be shared by animals? But if, said Darwin, by the imagination we mean that the mind puts together images and ideas to create new results, then dreaming must surely demonstrate imagination. And, 'as dogs, cats, horses, and probably all the higher animals, even birds have vivid dreams, and this is shown by their movements and the sounds uttered, we must admit that they possess some power of imagination.'

He focused on the howling of dogs at night. 'There must be something special,' he argued, that makes dogs howl, 'and especially during moonlight, in that remarkable and melancholy manner called baying.' For later editions of his book Darwin was happy to find a French expert, Houzeau: 'Houzeau thinks that their imaginations are disturbed by the vague outlines of the surrounding objects, and conjure up before them fantastic images: if this be so, their feelings may almost be called superstitious.'

This theory of Darwin's about dog 'superstition' developing in the

face of inexplicable natural forces has undertones from elsewhere. For early Victorian anthropologists, human religious belief first arose when primitive people were faced with mysterious events, such as earthquakes, storms or floods. Humans had a tendency, such anthropologists argued, to fear that such 'uncaused' events had a 'causer'. For Darwin to use the word 'superstition' in this context was daring; he was now treading dangerously close to implying that lower animals might possess something parallel to primitive religious beliefs.

At this point, Darwin went to sum up his argument so far:

> It has, I think, now been shown that man and the higher animals, especially the primates, have some few instincts in

> common. All have the same senses, intuitions, and sensations – similar passions, affections and emotions, even the more complex ones, such as jealousy, suspicion, emulation, gratitude, and magnanimity; they practise deceit and are revengeful; they are sometimes susceptible to ridicule, and even have a sense of humour; they feel wonder and curiosity; they possess the same faculties of imitation, attention, deliberation, choice, memory, imagination, the association of ideas, and reason, though in very different degrees ... Nevertheless, many authors have insisted that man is divided by an insuperable barrier from all the lower animals in his mental faculties.

So Darwin wasn't content; he pressed on. Having made his way through the emotional and intellectual life of animals, he began the final stage of his argument:

> It has been asserted that man alone is capable of progressive improvement; that he alone makes use of tools or fire, domesticates other animals, or possesses property; that no animal has the power of abstraction, or of forming general concepts, is self-conscious and comprehends itself; that no animal employs language; that man alone has a sense of beauty, is liable to caprice, has the feeling of gratitude, mystery etc; believes in God, or is endowed with a conscience.

He had reached the central assumptions made about the uniqueness of human beings. In this next section of his book, Darwin set out to tackle these assumptions, too.

On the last leg of his journey now, Darwin began to move into areas that must have seemed extraordinary to his readers. One of the most striking of these is animals' potential for abstract thought and self-consciousness. Abstract thought had always been seen as a kind of cornerstone of human originality, yet here's Darwin describing one dog meeting another on a walk:

> When a dog sees another dog at a distance, it is often clear that he perceives that it is a dog in the abstract; for when he gets nearer his whole manner changes, if the other dog be a friend.

For Darwin, if the dog recognises first the abstract concept of 'dog', but then shifts, then they are doing exactly the same as a person would do. So, he continued,

> In all such cases it is a pure assumption to assert that the mental act is not essentially of the same nature in the animal as in man. If either refers to what he perceives with his sense to a mental concept, then so do both.

Darwin then tells the story of Polly, his terrier:

> When I say to my terrier, in an eager voice (and I have made the trial many times), 'Hi, hi, where is it?' she at once takes it as a sign that something is to be hunted, and generally first looks quickly all around, and then rushes into the nearest thicket, to scent for any game, but finding nothing, she looks up into any neighbouring tree for a squirrel. Now do not these actions clearly show that she had in her mind a general idea or concept that some animal is to be discovered or hunted?

Darwin was not simply trying to make the point that dogs are clever companions; he was trying to prove that human distinctiveness is a myth, a story we have told ourselves to console ourselves for our humble origins. The point of the exercise always was to argue that human beings and the lower mammals exist in the closest of relations to one another. Darwin even tackled tool-making and property rights. Darwin cited the example of chimpanzees and elephants using tools; and as for the idea that only humans possess a concept of ownership of property? 'This idea is common to every dog with a bone,' he says, with what we can only imagine was a smile.

Darwin was particularly intrigued by the problem of whether an animal's actions could ever be called 'moral'. In the Victorian era it was generally believed that animals were guided by instinct, and humans acted through choice. But for Darwin, there was a profoundly grey area in between the extremes of an animal who was acting completely on instinct, and a human being who acts in a moral fashion after some reflection. For Darwin, some dogs could clearly be seen to be deliberating between two choices of action, weighing their instinct to rescue comrades from danger, against their desire for self-protection.

Darwin described watching a dog struggling between different instincts, trying to assess which it should follow: 'As when a dog rushes after a hare, is rebuked, pauses, hesitates, pursues again, or returns ashamed to his master; or as between the love of a female dog for her young puppies and for her master – for she may be seen to slink away to them, as if half ashamed of not accompanying her master.' If dogs could veer between two different decisions, if they understood they had two options to choose from, then surely their choosing the dangerous one was a moral choice?

Having made this point, he attacked the issue from a different perspective, taking the example of someone who dives into water to help a drowning stranger. In fact, argued Darwin, we are more likely to hold in high esteem the person who acts without thinking in this situation; the person who doesn't have to weigh up the pros and cons before diving in, the person who immediately understands the necessity in the situation. This is exactly what we assume the dog would do. Yet, said Darwin, 'when a Newfoundland dog drags a child out of the water, or a monkey faces danger to rescue its comrade, or takes charge of an orphan monkey, we do not call its conduct moral'. In conclusion, Darwin wrote, once more

BLOODHOUND.

concluding in favour of the animal world, dogs have 'something very like a conscience'.

Darwin had concluded in favour of the moral sense of animals; now he came to self-consciousness and language. Even today, self-consciousness in animals is a riddle, with some scientists choosing to use the term 'self-recognition' because it describes a slightly simpler version of being aware of the self. Yet Darwin, who had spent so many years observing animals, didn't hesitate. Of course, Darwin conceded, we can agree that a dog did not spend time pondering the question of the afterlife, the immortal soul, 'or what is life and death, and so forth.'

> But how can we feel sure that an old dog with an excellent memory and some power of imagination, as shown in his dreams, never reflects on his past pleasures or pains in the chase? And this would be a form of self-consciousness.

Darwin did not choose to argue this case with great certainty; he simply offers us a thought experiment, and the question, how can we feel sure that it never happens? Then Darwin turned to language, normally seen as 'one of the chief distinctions' between man and the lower animals. But language as Darwin defined it, was a more open field: language expresses what is in a creature's mind, and allows that creature to understand what is expressed by another.

Dogs certainly understand what people say to them, Darwin argued. 'As everyone knows,' he said, 'dogs understand many words and sentences'. So the distinction between man and lower animals can't be defined as the ability to comprehend speech. For Darwin, dogs were just like babies 'between the ages of ten and twelve months'; they 'understand many words and short sentences, but cannot yet utter a single word.'

But dogs can communicate in return, says Darwin; a dog's bark can do just that. As he lists:

> We have the bark of eagerness, as in the chase; that of anger, as well as growling; the yelp or howl of despair, as when shut up; the baying at night; the bark of joy, as when starting on a walk with his master; and the very distinct

> one of demand or supplication, as when wishing for a door or window to be opened.

Darwin spent more than a hundred pages of *The Descent of Man* discussing those 'lower animals', and the book is full of charming and thought-provoking examples. But given the circumstances of Darwin's writing, there is one passage in *The Descent of Man* which continues to deliver a shock on reading, even now. It is difficult to imagine how a Victorian reader might have felt when Darwin returned to the issue of animals and what they think about the causes of natural events, aiming to throw some light on the development of religious feelings.

For Darwin, this is not a question of belief in a single, Christian God. This is religion defined as 'the belief in unseen or spiritual agencies'. Darwin followed his contemporary, the anthropologist E. B. Tylor, who had argued that the belief in the supernatural may have come about initially through the process of dreaming, as human beings first came to believe in spirits which inhabit natural objects. But Darwin also believed in an even 'earlier and ruder stage', when individuals are inclined to ascribe life, intention and mental powers that are similar to our own, to anything which has power or movement.

For Darwin, this allowed him to connect people he saw as 'savages' with animals:

> My dog, a full-grown and very sensible animal, was lying on the lawn during a hot and still day; but at a little distance a slight breeze occasionally moved an open parasol, which would have been wholly disregarded by the dog had any

> one stood near it. As it was, every time that the parasol slightly moved, the dog growled fiercely and barked. He must, I think, have reasoned to himself in a rapid and unconscious manner, that movement without any apparent cause indicated the presence of some strange living agent, and that no stranger had a right to be on his territory.

For Darwin to make a parallel between this dog's fear of the parasol, and early human beings' development of religious belief, seems an extraordinary step. Yet he took it. The dog barks at the umbrella because it thinks it is being moved by a 'strange living agent'. 'The belief in spiritual agencies,' he wrote, describing the process he believed took place amongst 'savages', 'would easily pass into the belief in the existence of one or more gods.' Watching his dog, barking at an umbrella, in other words, allowed Darwin to see how belief in God might come about. It is difficult to imagine how this suggestion failed to cause more offence, except for the gentleness in which the idea is couched, on an English lawn, on a lazily hot summer's day. He continued:

> The feeling of religious devotion is a highly complex one, consisting of love, complete submission to an exalted and mysterious superior, a strong sense of dependence, fear, reverence, gratitude, hope for the future, and perhaps other elements. No being could experience so complex an emotion until advanced in his intellectual and moral faculties to at least a moderately high level. Nevertheless, we see some distant approach to this state of mind in the deep love of a dog for his master, associated with complete submission, some fear, and perhaps other feelings.

Although Darwin says the resemblance is 'distant', he nonetheless believes it exists: religious devotion is something like the love a dog has for his master. And if the picture weren't complete, he adds, finally: 'Professor Braubach goes so far as to maintain that a dog looks on his master as on a god.' In the conclusion to the book Darwin certainly concedes that the distance between the lowest human, and the highest animal, is 'immense'. But, he adds, 'the difference ... great as it is, certainly is one of degree and not of kind.' There is an underlying implication here: when a human being believes in God, they are a little bit like a dog looking up to its master.

What Darwin achieved in *The Descent of Man* could be viewed in a number of ways. For some, *The Descent of Man* is a lesser book than the *Origin*, for it dilutes the strong force of the idea of natural selection. For other observers, it is all about humanity, about races and slavery, written right in the heart years of the American Civil War. Yet it is hard, when reading the first half of the book especially, to ignore the insistent suggestion that the book is really about how human beings are no better than their pets. Or perhaps not quite that way round. Perhaps the best way of summing up Darwin's argument is that our pets are at the very least as good as us.

The Descent of Man showed enough of Darwin's love for his dogs to attract new fans. In the course of research, Darwin became friends with George Cupples, a breeder of deerhounds. After publication, he exchanged letters with Hugh Dalziel, a newspaper critic of dog shows.

And despite all the speculative remarks made about religious belief in the text of the book, Darwin also got a delightful letter of fan mail from an old friend: Down's former vicar, John Brodie Innes.

He wrote from his Scottish home to say how much he had enjoyed reading the *Descent,* though it had not converted him: 'I hold to the old belief that a man was made a man', he wrote.

Innes had been the vicar at Down for many years, and was a proper Church of England clergyman with entirely orthodox views. Though Innes believed that God had created the world and Darwin did not, they always had a warm relationship. This may be because, despite his ungodly writings, in practice Darwin was a relatively helpful congregation member, subscribing to five local Sunday Schools and running the Down Coal and Clothing Club, a savings club whose funds were dramatically augmented by the local gentry. Darwin spent years trying to find Innes a house in the village whilst he still worked there, and later in trying to ensure the church had a good curate (who didn't 'walk about with girls at night'). Darwin even made sure that the organ got repaired.

They were also personal friends: when Innes had pain in his jaw, Darwin sent Innes some Arnica for his toothache. Darwin sent references about bees; Innes returned with information about pigeons. Darwin expressed himself sceptical about Innes' reports of wheat growing from oats, toads appearing from inside rocks, a cow that had been crossed with a red deer, 'land barnacles' that turned out to be lichens, and 'spirit-rapping'; Innes sent his congratulations when Darwin's son George distinguished himself at Cambridge.

Most importantly, they shared a liking for dogs. Innes had 'various animals and pets' according to one of Darwin's letters. When Innes read *The Variation of Plants and Animals Under Domestication* he sent Darwin an example of a pointer that confirmed Darwin's argument. When he read *The Descent of Man,* he sent Darwin Brodie family dog stories that corresponded to Darwin's argument. And when Innes' family moved to Scotland, the Darwins took in their pet dogs Tartar and Quiz, who were fostered at Down House.

Darwin wrote to say: 'we shall be delighted to have Quiz, who shall be taken great care of, and never parted with, and when old and infirm shall pass from this life easily.' Many letters were exchanged negotiating the transfer, and once Quiz had finally arrived, Darwin wrote a letter that gave as much detail as if he had one of Innes' children in his care: 'I am heartily glad to say that Quiz arrived last night safe and sound (but with a cough) and has been running about the house quite happy and very polite to every human being, including cats.'

Whatever he wrote in his books, Darwin, the evolutionist, and John Innes, who believed in the creation of the world by God, were able to be good friends. 'Certainly you and I never were like to quarrel over our differences, thanks mostly to your most kind forbearance with some hot headedness etc ...', wrote Innes in 1871.

Innes always made it clear that he was not convinced by Darwin's theorising, but it was a matter of pride that they didn't fall out over it. 'I have an abhorrence of an ape,' Innes wrote in May 1871, having just finished reading the *Descent,* 'but in my boy days had a very favorite little ring-tailed monkey, and I should much prefer one of that kind as my more immediate ancestor. Please think of my request favourably.' And at the end of the letter he added, with great lightness: 'I think on consideration I had almost rather have a dog for an ancestor than even a ring-tailed monkey.'

Darwin managed an equally pro-canine reply: 'Thanks for the very curious story about the dog and mutton chops. They are wonderful animals, and deserve to be loved with all one's heart, even when they do steal mutton-chops.' The criticisms of Darwin's book were simply and politely passed over. Whatever their disagreements, these two men agreed on one thing. As Innes once wrote to Darwin, with great warmth: 'How nicely things would go on if other folk were like Darwin and Brodie Innes!'

DOGS WATCHING FOR RATS.

Animals whom we have made our slaves we do not like to consider our equals.

After the publication of the *Descent,* Darwin spent the last decade of his life peacefully at home, receiving occasional visitors (Joseph Hooker, now Director of Kew, was never turned away) and continuing to experiment, write and publish. He worked in his greenhouse, carrying on many practical investigations, verging on the eccentric when it came to his study of garden worms. He even roped in his son Francis to play the bassoon to some of them, to test how they responded to vibration and sound.

Darwin's health was very poor in the last years of his life, though he still managed to finish writing important books on carnivorous plants, on cross-fertilisation in plants, and on those earthworms. One tragic event managed to have a silver lining: in 1874, Darwin's son Francis had married, and two years later the Darwins' first grandchild was born at Down House. Sadly Amy, the baby's mother, didn't survive the birth; her husband was distraught, and Bernard, the baby, and Francis both lived at Down from then on. Francis helped with secretarial work that may have been a useful distraction from his own grief, and Charles and Emma loved having their son and his little boy around.

Bernard grew up to become a golfing writer, and in one volume of memoirs he recalled his childhood, where he would play golf on his own: 'the easiest of games at which to play in pretending matches by oneself.' He also revealed his thoroughly typical Darwin method of naming his imaginary golfing pals: 'The names of the players were invented by taking the names of all dogs, cats and horses, whether our own or our relations, and adding to them either prefixes or suffixes. I could repeat many of them now.'

During these years Emma wrote often to her children giving news

IRISH WOLFHOUND.

of the dogs. The letter recalls the ones sent to Darwin by his sisters while he voyaged on the *Beagle:* Emma Darwin carried on the family tradition, thoroughly concerned with the dogs as members of the immediate circle. When Bob became ill one spring, Emma wrote to Henrietta: 'Poor Bobby is better today and has eaten a little. He looked so human, lying under a coat with his head on a pillow, and one just perceived the coat move a little over the tail if you spoke to him.'

Emma wrote to her children with advice on their own pets, too. When William, her eldest son, acquired a new puppy that failed to settle, she wrote to her daughter-in-law: 'I hope your poor puppy will soon be happier. Would not Otter [their other dog] sleep with him?' And she and Charles tried repeatedly to rehome

Pepper, the badly behaved dog of their son George. Pepper had the unwelcome habit of biting gardeners, and Emma worried that he would have to be put down as a result. 'I am vexed about Pepper. I feel it quite sad to extinguish such a quality of enjoyment as lived in that little body.' Pepper was eventually reprieved; after failing to settle in Leslie Stephen's household he went to live with the Archbishop of Canterbury.

It was Polly, though, the terrier who had belonged first to Henrietta, who took up the most inches in letters. Her distinctive character fascinated Emma: 'I never saw such a methodical dog. She sits on the mat when we go to lunch, to wait for her dinner, and on the rug in the chair by the stove when we go into dinner.' After losing a litter of puppies, Polly became fixated on Francis, and Emma wrote: 'I think she has taken it into her head that Francis is a very big puppy. She lies upon him whenever she can and licks his hands so constantly as to be quite troublesome.' Darwin was terribly fond of Polly and taught her to catch biscuits off her nose, and she sat with him patiently while he spent quiet hours working in his study.

But finally in the spring of 1882 Darwin's condition grew worse. In the middle of April he became so poorly that his family knew the worst had come. Henrietta began to take detailed notes for an account of his death, so that it could never be alleged Darwin had repented and called for God's help in his final moments. After his agonising last hours, held by Emma, he died during the night of the eighteenth.

The next morning, Bernard, Darwin's little grandson, and his father Francis walked down the garden one last time and picked wild arums, 'Lords and Ladies', in memory. But the Darwin family's hope of a simple burial next to his brother Erasmus was not to be. 'It gave us all a pang not to have him rest quietly by Eras', Emma said, but Darwin's champions had arranged for nothing less than Westminster Abbey. Polly, Darwin's favourite dog, survived her master by only a few days; she, however, was buried at Down, under the garden's apple tree.

Darwin died while there was still a great deal of uncertainty about his theory of descent with modification by natural selection. Even some distinguished scientists felt Darwin's case remained unproven. He himself added provisos to later editions of his books, qualifying the idea that natural selection was the sole mechanism of evolutionary change. So if Darwin were suddenly back with us today there would be hundreds of important questions he would want answered. Turning up in the Darwin Wing of the Natural History Museum, or at his old college, Christ's in Cambridge, he would be full of queries, objections and possibilities.

We know this because he expressed his frustration with the gaps in his knowledge within the text of his most famous book, *The Origin of Species.* So he'd surely be delighted to know how many of the gaps that he identified we've now managed to fill. Particularly interesting to him would be the fossils of human beings' nearest relatives, dating back over the last two million years in Africa. We can now tell almost the whole history of our ancestors as they gradually walked more upright, began to make tools and developed fine speech. And knowing what we now do about Darwin's interests, we can be sure that for a start he'd love to know the latest about the evolution of the domestic dog.

Darwin had always been fascinated by the problem of how dogs came to be domesticated; after all, he knew that animals like our pet dogs don't exist in the wild. Like many other naturalists before him, he believed that dogs had been tamed by human beings from wild ancestors. This domestication must have happened a long time in the past: nineteenth-century archaeologists had found Neolithic and Bronze Age sites where human skeletons were buried in close association with dog bones. These dogs were not very wolf-like: they weighed much less than a wolf, and had the distinctive puppy-like 'faces' of domestic dogs, quite unlike the more pointed muzzles of wolves.

And by about four or five thousand years ago, Darwin observed, many distinct breeds were recognisable: a mastiff-likc guard dog, appearing in giant relief on a stone panel from ancient Babylon; greyhound–like hunting dogs, with long legs and keen eyes, pictured in Egyptian tomb paintings; and in Roman times, the first lap dogs and turnspits, taking the edge off the barbarity of life, warming cold hands and roasting the dinner. If these dogs were domesticated, they had been domesticated a long time ago, many thousands of years in the past, and had stayed roughly the same ever since. But where had they originated?

After much reflection, Darwin concluded that dogs were descended from a number of different ancestors. When he looked at dogs in all their variety, from bull mastiffs to tiny spaniels, he felt there was too much difference for them all to be bred from a single wild species. The different traits of different breeds, said Darwin, must have come from a variety of different wild ancestors, hybridised together. 'Two good species of wolf, *Canis lupus* and *C. latrans,*' he began, as if he were writing a recipe, 'and from two or three other doubtful species of wolf; from at least one or two South American canine species; from several races or species of the jackal; and perhaps from one or more extinct species.'

Darwin could only hypothesize about the ancestry of the dog. He didn't know about genetics, or the existence of DNA; in Victorian times, it wasn't yet understood how an individual passes its characteristics on to the next generation. Today, we understand the process in great detail, and we can study it in the lab. And in the last twenty years it has become straightforward and cheap to compare sections of DNA from many different individuals or species. As a result, we can sequence a dog's DNA, or 'genome', and compare it to a wolf's, or a coyote's, to see how closely they're related.

Using these kinds of comparison techniques, we can also estimate the amount of evolutionary time since two groups of animals separated from each other. By doing so, science has developed a way of plotting the evolutionary relationships of the dog family on a branching tree; a tree that looks very much like that very first diagram Darwin drew, way back in 1838.

We now know that all domestic dogs, from chihuahuas to Great Danes, as well as wolves, dingos, dholes and foxes, are part of one distinct family called the 'canids', a lineage in the group of mammals called the carnivores. Dogs, foxes and wolves split from the other carnivores, such as bears and big cats, about fifty million years ago. This makes them a relatively ancient lineage, and many species in the group are now extinct. Today, there are around thirty-five living species of canid which all evolved fairly recently, in a burst of evolutionary activity around twelve to fifteen million years ago.

POODLE.

One of these canids is a distant relative of the wolves and dogs, a rare South American creature called a 'zorro', discovered by Darwin himself during the *Beagle* voyage. Known only on an island off the coast of Chile, it is named after him, 'Darwin's fox'.

Today, there are many scientists around the world working on the dog genome. The first dog to have some of its genome mapped was a poodle called Shadow, in 2003. Not coincidentally, Shadow was the pet of biotech millionaire Craig Ventner, the first human being to have his DNA analysed in full. The initial map of Shadow's DNA revealed some intriguing findings: the poodle shared an astonishing 18,473 genes with human beings, out of the human total of 24,567. It turns out that dogs and human beings have a good deal in

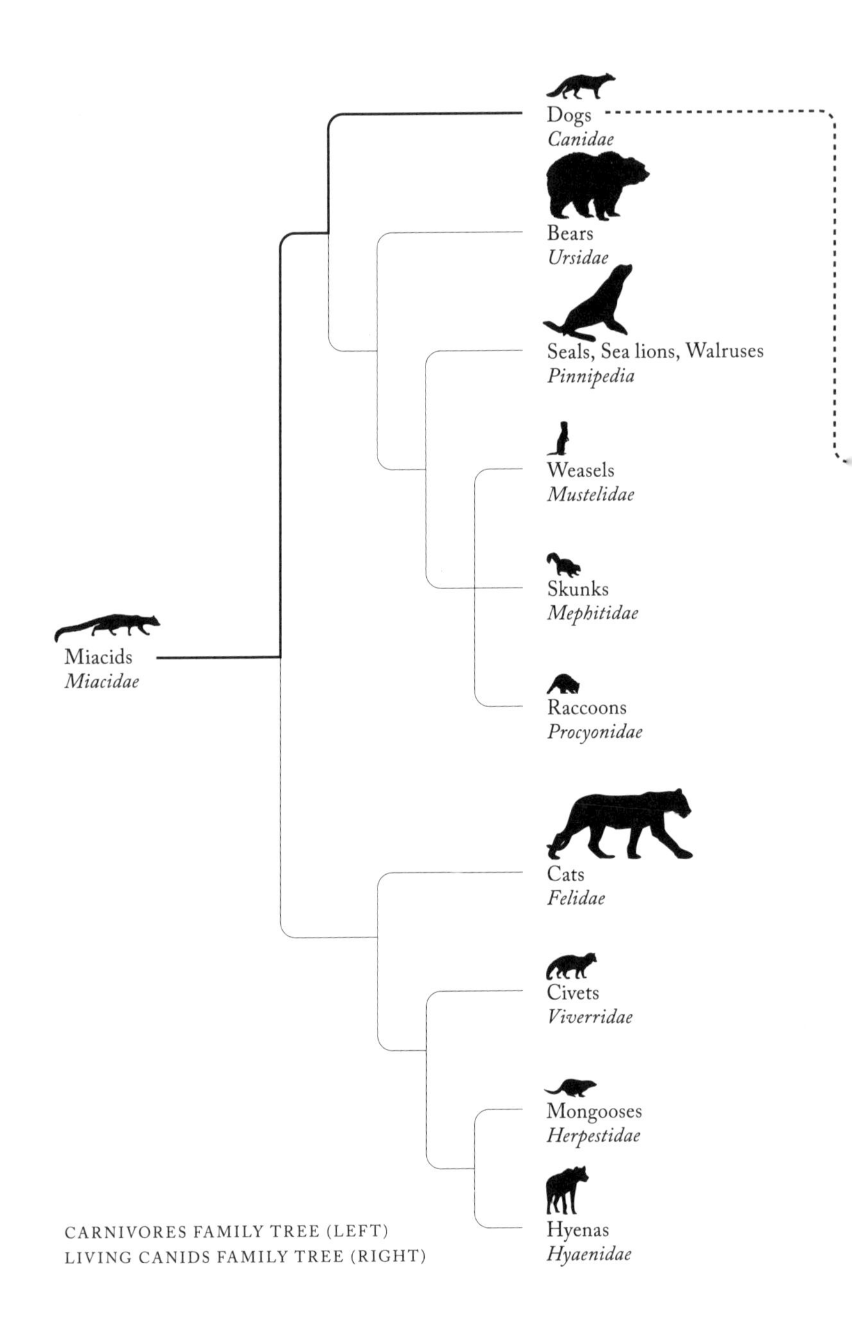

CARNIVORES FAMILY TREE (LEFT)
LIVING CANIDS FAMILY TREE (RIGHT)

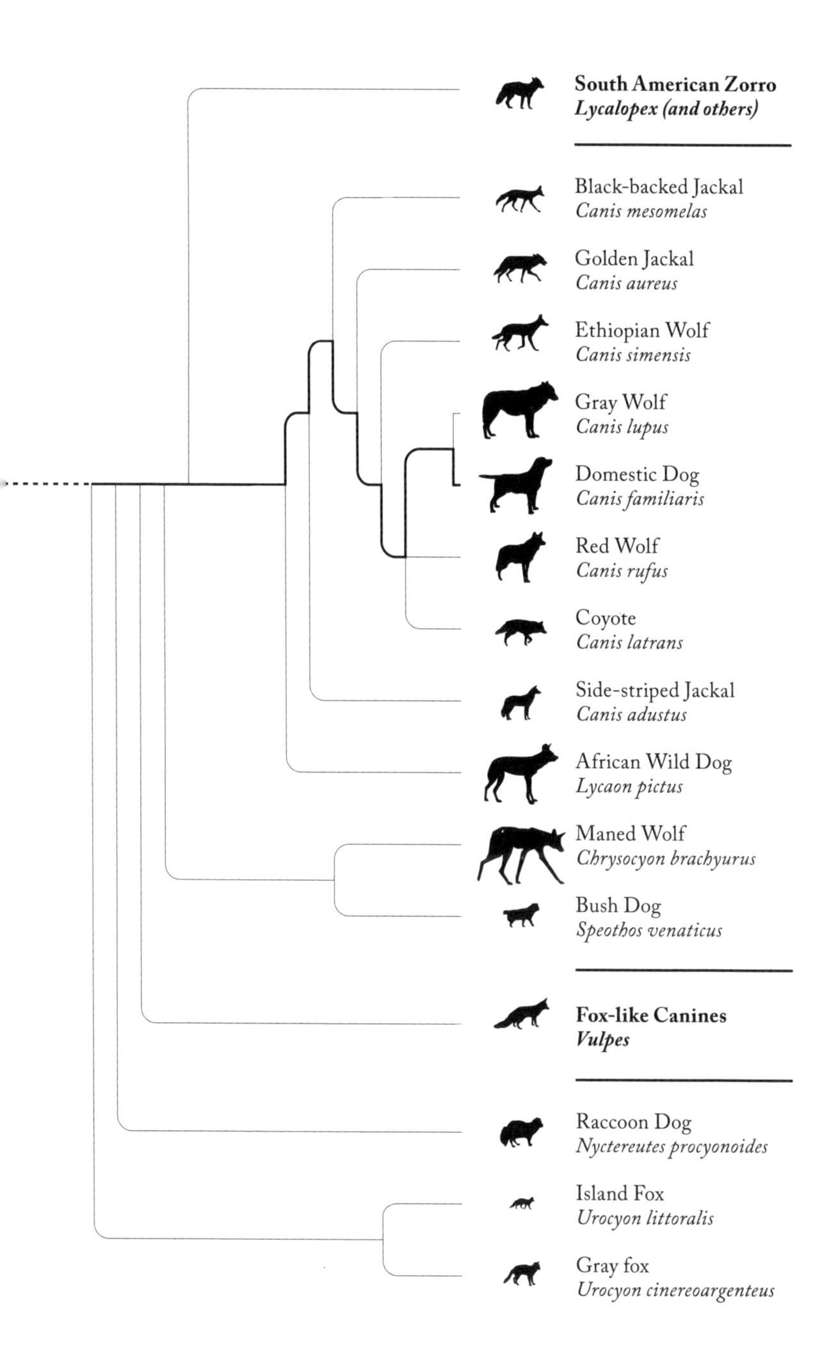
South American Zorro
Lycalopex (and others)
Black-backed Jackal
Canis mesomelas
Golden Jackal
Canis aureus
Ethiopian Wolf
Canis simensis
Gray Wolf
Canis lupus
Domestic Dog
Canis familiaris
Red Wolf
Canis rufus
Coyote
Canis latrans
Side-striped Jackal
Canis adustus
African Wild Dog
Lycaon pictus
Maned Wolf
Chrysocyon brachyurus
Bush Dog
Speothos venaticus
Fox-like Canines
Vulpes
Raccoon Dog
Nyctereutes procyonoides
Island Fox
Urocyon littoralis
Gray fox
Urocyon cinereoargenteus

common at the genetic level (which may bring a smile to the face of those who like to think that dogs and their owners are often rather alike).

Then, in 2006, scientists completed the first full map of the genome of a boxer dog called Tasha. This was important because a genome is composed of useful DNA, like genes, but also junkier DNA, such as unnecessary repeats. Though 'junk DNA' might not sound worthy of academic study, over time it's been learned that it is often more useful than the genes themselves for looking at evolutionary relationships. Because it's junk, and doesn't get used, it's often a better record of the past: a bit like a crammed family attic full of old papers might be more informative about the history of a house's inhabitants than their clean modern kitchen.

So using these new techniques, what did scientists find out about the ancestry of dogs? Well, Darwin would be pleased to find out that the DNA studies confirm his suspicions: dogs and wolves are extremely closely related. The average difference between the genomes of dogs and wolves was about 1%; that is compared to the next closest relation, a 7.5% divergence between dogs and coyotes.

However, this doesn't mean dogs are simply tame wolves. Though some fashionable dog trainers would have you believe otherwise (and in the US the dog is still classified as part of the wolf species *Canis lupus*) it is more accurate to speak of dogs having a 'wolf-like ancestor'. Wolves are dogs' closest living relative, but we cannot say more than this: the original wolf-like ancestor may have died out as dogs were domesticated and came to live within human communities.

Dog genome experts had one more intriguing finding. The dog DNA they studied could be divided into four major groups, each of which was descended from one ancestor. Each group represented a single 'domestication event'. So dogs were domesticated from a wild, wolf-like ancestor more than once, but probably just a few times overall. DNA evidence also suggests that this initial taming of a wild ancestor took place more than a hundred thousand years ago. So human beings had their own breeds of tame canids which no longer regularly bred with wild wolves, while they were still living a hunter-gatherer lifestyle, long before settling in village communities.

Yet this picture of stone age pet owners left wolf and dog behaviour experts skeptical. Whatever the genetic evidence, they wanted to understand how prehistoric people had tamed wolf puppies to become useful members of a community. Wolves are greedy and possessive about food, and a tame wolf would have been more likely to hinder than help on hunting expeditions. In addition, wolves aren't good guards; they will run away from their den if disturbed, rather than raising an alarm. How then, asked these critics, could the domestication of those wolf-like ancestors actually have come about?

Now, the distinction first made by Darwin between natural and artificial selection came back to centre-stage. All along it had been assumed that human beings adopted wolf-like puppies from the wild and raised them, gradually breeding in more tolerable traits: artificial selection in action.

But what if the first dogs became tamer because of natural selection? These wolf-like ancestors weren't welcomed into human communities; their presence at the edge of camping grounds was merely tolerated. Bones of wolves have been found in association with sites of human settlement as far back as four hundred thousand years ago, and perhaps they were tamer than the average pack, living on the margins of human communities and scavenging amongst human rubbish. The more submissive and tolerant the wolf-like ancestor was, the easier it would be for that animal to find food from scraps discarded and hunting remains. The less frightened it was of human beings, the less energy it would use up running away unnecessarily every time someone came close. And, thus, the easier it would be for that individual to survive and reproduce.

If wolf-like ancestors initially lived alongside human communities without being fully integrated into them, there would have been a two-stage process of domestication. In the first, the wolf-like ancestors began living alongside the humans, scavenging from their waste piles. In the second, the wolf-like ancestors began to be tamed, and were cared for within the human community, rather than living on its margins. Today, most dog experts agree that something like this two-stage process led to the domestication of the dog.

There is one more piece of intriguing evidence to add to the story. In the late 1940s, a Russian scientist called Dimitri Belayev began studying silver foxes, hoping to produce animals that were happier to be raised in captivity for their fur.

Belayev selected through several generations of foxes, each time only breeding from the tamest animals. Within eight generations, he began to produce much more sociable animals, which were far more tolerant of human beings. But his research also had some other rather unexpected results: the tame foxes had patchy coloured coats, floppy ears, rolled tails, and smaller skulls. In taming the original wild, aggressive foxes, they had also become more 'dog-like'.

Belayev's study suggests that from tameness to floppy ears, all these characteristics are associated with a 'linkage', meaning they come together as a bundle. The 'linkage' effect is not unique to dogs. Across the whole field of genetics, these linked genes have been found to be inherited together because they are located in a similar place on the chromosome.

By selecting for one characteristic, a breeder may unintentionally end up getting all the others too. If natural selection favoured the wolf-like canids which were tamer and more readily tolerant of human activity, it's possible that these animals were more dog-like in appearance, too, due to this genetic linkage. In other words, human breeders may have had less to do with the floppy ears and patchy coats of dogs than we previously thought; it's simply a happy accident.

Reading all this, you may have wondered how scientists have been able to justify spending so much time on the esoteric question of how the dog evolved. But rather surprisingly this work has many useful applications. One of the most significant findings is that there's a lot of similarity between the human genome and that of the dog, for this coincidence provides some remarkable opportunities for medical science.

Dogs and human beings are both highly prone to genetic diseases, illnesses caused by flaws in our DNA. Over three hundred and fifty different genetic ailments occur in dogs, all of which are much more common in pedigree animals, and many of these diseases have parallels amongst human beings. The highly inbred nature of many pedigree dogs has been much discussed in recent years, leading eventually to the controversial 2009 decision by the BBC not to screen the dog show Crufts on terrestrial television. But the suffering of some inbred pedigree animals might at least render one useful outcome. Pedigree dogs are so inbred, it turns out, that they provide a really efficient way of looking at how genetic disease comes about.

From retinal disease in miniature schnauzers, to hip dysplasia in greyhounds, pedigree dogs are plagued by genetic diseases, being much more likely to succumb to diseases with a strong genetic component, such as cancer and diabetes, too. Some illnesses, such as the rare degenerative Lou Gerig's disease, occur in exactly the same form, caused by the same gene, in German shepherd dogs and human beings. Battens disease, a terrible inherited neurological illness in humans, also affects Tibetan terriers. Dobermans suffer from narcolepsy, and dogs with the condition can be put to sleep by the sound of a single clap.

But the study of genetic disease in human beings has always been hampered because the human genome has a high degree of variation, and pinning down the cause of a disease to a single gene or group of genes is often incredibly time-consuming. In contrast, pedigree dogs, with their generations of in-breeding, have a shockingly narrow 'gene pool'. For genetic diseases in dogs, in-breeding often means that a small number of genes, or even just one, can be quickly identified as causing the illness. And pedigrees provide a unique source of family information for tracing the development of a malady, too.

So the dog genome offers particular hope to human sufferers of genetic diseases. It provides possibilities for the development of treatments, and for devising tests to let people know if they are carriers of a faulty gene. And with luck, it offers future hope to canine sufferers, too.

* * *

Today, we have a good picture of the process of canine evolution, which Darwin first theorised about in the mid-nineteenth century. But what does modern science have to say about the ideas Darwin put forward in *The Descent of Man*? In that book, as we have seen, Darwin made many claims for the interior life of animals. But can we really agree with Darwin that dogs love, that they imagine, and most outlandishly, that they have a sense of humour?

For much of the twentieth century Darwin's picture of dogs was disregarded. During those years, a theory of animal behaviour called 'behaviourism' came to the fore. Its proponents said that human beings should not assume anything about animals based on inference from their own experience. Animals didn't have awareness, or memories, or emotional attachments: to the behaviourists, animals were simply complex machines, acting on instinct.

In other words, Darwin was making an unjustified assumption when he argued what a dog felt when it looked up to its master. Darwin was not free to do this, argued the behaviourists. It was 'anthropomorphism' – attributing human thoughts and feelings to an animal that was not able to explain its feelings for itself. As we don't have good information about what an animal thinks and feels, they argued, we shouldn't make any assumptions about it at all.

Things have shifted since the heyday of behaviourism. But for many animal behaviourists, even today, this is still the only tenable position. For them, it is also the only rigorously scientific approach, as it doesn't require making any unjustified leaps of faith. You cannot test the interior thoughts of an animal: you must therefore disregard them experimentally.

Yet anthropomorphising has its uses. One can proceed 'as if' a dog is loyal. And in our everyday dealings with animals, we do exactly this. We anthropomorphise when we trust a dog to be left in a room with a baby, thinking that it is 'loyal', or when we decide not to do so, because it is prone to be jealous. We behave 'as if' our dogs love us, or 'as if' they are capable of having a general concept of pheasants that will always make them disappear in full pursuit.

And there have always been a minority of naturalists studying animals who felt that the arguments of the behaviourists contained a kind of fallacy. The impetus has come from ethologists who advocate studying animal behaviour in its own context. Jane Goodall, the famous chimpanzee specialist, is amongst them. For Goodall, there is little doubt that animals experience some of the same emotional states as we do, and Darwin was right to say so.

But if there is one thing that comes through reading all of his writings, it is that Darwin believed in evidence. He believed in collecting, sifting, assimilating and testing evidence. It is not enough to say the dog is behaving as if it is loyal: this must be tested. So today, many animal behaviourists, particularly those working with primates, are preoccupied with devising testable hypotheses about the complex interior lives of higher mammals – from their metaphysics, in the case of Dorothy Cheney and Robert Seyfarth's baboons, to their moral choices, in the case of Franz de Waal's chimpanzees. Marian Stamp Dawkins, Professor of Animal Behaviour at Oxford, even interests herself in the emotional lives of farmed chickens.

This growing field is called 'cognitive psychology'. Today, its aim is to look at the 'mental states' of sophisticated mammals such as dogs. Its proponents explore the maps of the world which dogs hold inside their heads; they test the memory and problem-solving skills of dogs; most of all they try to work out which of the many claims made by dog owners of their pets actually turn out to be true.

For cognitive psychologists, there is plenty of reason to think that Darwin was right to say animals experience emotion, and that they have a measure of self-awareness. Most importantly, emotion and self-awareness must have adaptive functions. For example, simple self-awareness allows an animal to make decisions about what is causing a movement in the world around it, and about how to react. A more complex self-awareness lets monkeys predict what other monkeys can see from their vantage point, allowing scientists to see mischievous deceptions taking place.

Intriguingly, academics even have an experiment that helps to illuminate Darwin's little suggestion that dogs possess a sense of humour. Two animal psychologists called Robert Mitchell and Nicholas Thompson decided to test dogs and humans for deception. They set up a test which directly corresponds to the example Darwin described in *The Descent of Man;* the dog runs at the human being, veering aside at the last minute. Or, the dog gives up a ball, and then snatches for it just as the human being reaches out for it. Dogs really like doing this: 92% of the dogs tested chose to do it.

And it turns out that deceiving is a particularly significant thing for a dog to be able to do, because it means that the dog can conceive of another consciousness and make simple predictions about what the other consciousness is going to do. For scientists from the new cognitive psychology school, complex mental states such as trust, humour and deception can really exist in animals.

And it's that trust between a human being and a dog which tells us something important about how dogs came to be. Approximately fifteen thousand years ago, human beings went from allowing wild canid ancestor to graze their trash, to adopting puppies and raising them by their own hearths. Sometimes, in times of want, they ended up eating the puppies. Mostly, they valued them. From the middle stone age, dogs were buried alongside their owner, a life's companion going out into the hereafter, a protector for the unknown dangers of an unknown journey.

Dogs have been shaped by the breeding desires of their owners. But there have also been more unconscious processes at work. The most appealing puppy was most likely to get fed; it may not necessarily have been the toughest or the fastest, it may just have been the one with the cutest face.

Dogs evolved to work on our unconscious as well as our conscious desires. We feel affection for dogs because we feel we are loved in return, but many skeptics would argue that selection pressures have acted to give the best chances of reproductive survival to dogs

OLD-FASHIONED ENGLISH SETTER,—RETRIEVERS, ONE A CROSS WITH BLOODHOUND.

that make us feel loved. The dog we feel the most loved by, we are most likely to dote on, to protect, to feed and keep warm, to defend from danger.

Despite this cynical view, most of us nonetheless love our dogs for themselves. We love dogs because of their gentle way of fitting into our lives, of generally not minding too much when a walk gets skipped because it is pouring with rain; of licking your face when you are crying; or of tolerating, though they haven't the faintest idea why you are making them do it, posing for a photo.

At bottom we make the assumption that the dog is a little bit like us. We anthropomorphise. We say a dog is depressed because its owner rehomed it, or that it's worried because it can tell we are packing for a holiday. We see a dog lying down, twitching, sniffing, yapping in its sleep, and we believe that it is dreaming.

How can you claim to know something like this? asks the hardest of the hard scientists. I trust my instinct, and why do you claim to know differently, unless you can prove otherwise?, counters the dog owner. Neither side can win; both stand opposite each other in a kind of stand-off. The truth is that we still don't have a deeply satisfactory answer to the question of whether dogs love, dream, or play jokes. But science is beginning to find ways of testing these questions, in ways that would have been profoundly intriguing to Darwin. In future, it's exciting and funny to think that one day we may know exactly what our pets are thinking.

On a deeper level, though, the debate always boils down to the same question: are animals like us, or not like us? Darwin was always willing to argue that animals are like us. That they feel happiness, sadness, grief and joy in some of the same ways that we do. That the differences between us are ones of degree, not of substance. Animals can reason, just not as well as we can. Animals can think and plan, and animals can love, just not in quite the same wordily self-conscious way that we do. Fundamentally, at root, Darwin said, looking at his dogs, we are animals, absolutely just like them.

Charles Darwin was one of the great observational naturalists, a man who watched the world, who always paid attention. And his love for his dogs was a passion that spanned his whole lifetime. In the last book Darwin ever wrote on the subject of higher animals, *The Expression of the Emotions*, Polly, his pet terrier appears almost all the way through the text.

Darwin's intention in the book was to prove that human express-ions were directly related to identical communication gestures in animals. It was yet another way of showing that humans and animals were all part of the same world. He aimed to construct an argument that applies to all living beings, but in many ways *The Expression of the Emotions* is the most personal of his books. The domestic companions of his life had never been present in quite such intimate detail. Polly is refered to as 'my terrier' throughout the book. And alongside Polly are Darwin's children, whose first smiles as babies are recorded in a chapter entitled, appropriately, 'Joy, High Spirits, Love.'

The Expression of the Emotions is a book that records Darwin's lifelong experience of owning dogs. Though many of the examples are drawn from Polly, he is recording behaviours that are familiar to him from all the pets he ever owned. And each behaviour he saw in one of his own pets had its significance within a wider picture of animal evolution. For Darwin, dogs were a vital key to his life's work. They made him ponder how adaptation worked; their own history posed questions about how species formed; and most of all, they were a keen reminder, every day, of the profound connection between the human and animal worlds.

It is sad to think that of all the dogs discussed in this book, we have images of only two. But if we began with Bob, let's end with Polly, the terrier. Here is Polly, immortalized forever, in one of the engravings for Darwin's book on how emotions are expressed in humans and animals. The little terrier stands with one leg raised, her head slightly tilted, her ears raised; she is clearly intrigued by what she sees. The plate was prepared from a photograph, though the original is no longer known.

Darwin uses Polly as the example, but once more, he evokes a whole lifetime of dogs:

> Dogs of all kinds when intently watching and slowly approaching their prey, frequently keep one of their fore-legs doubled up for a long time, ready for the next cautious step; and this is eminently characteristic of the pointer. But from habit they behave in exactly the same manner whenever their attention is aroused.

Throughout his life, Darwin balanced a sense of dogs as individuals, with their own personalities and drives, with a curiosity about how they worked as a species. And whilst sometimes his thoughts about dogs were very complex, he also never lost his gentle sense of what made them tick. The tenderness with which he portrays Polly takes us back to those early family letters, exchanged when Darwin took his first journey away from home as a teenager. So though Darwin's careful labelling of the image on the surface gives away very little about the depth of his feelings for Polly, it cannot help but give the reader a final smile:

> Small dog watching a cat on a table.

This book owes its existence to Jim Secord, with whom it was my great privilege to study for four years, first at Imperial and then in Cambridge. He wrote the piece on Darwin and the breeders which first drew my attention to the subject, but he was also unfailingly enthusiastic and encouraging in all the years he taught me and beyond. I owe him the greatest and most unrepayable debt of thanks.

The idea of the book began as a running joke during a summer school I taught on Darwin during the summer of 2006, and I would particularly like to thank that group of students for the congenial atmosphere that made thinking about Darwin in a slightly different way possible. I would also like to thank Darren Naish, Catherine Hall, Mark Cocker, Stephen Moss, Dominique Pourtaut-Darriet, Andrea Stuart, Julie Wheelwright, Jim Gill, David Roberts, Andrew Dalkin, Jim Endersby, Lena Corner, Tim Lewis, Bill Tuckey, Kate Burt, Jo Feldman, Pamela Neville-Sington, Jurgen Wolff, Sue Seddon, Christina Harrison, John Nicoll, Andrew Dunn and Nicki Davis, all of whom in various ways helped me straighten out my ideas; all my Darwin students, from 1994 to the present day; and all my friends at Kew, but in particular Dr Mark Nesbitt for some timely help on domestication.

For more detailed reading of the manuscript I am deeply grateful to Karen Townshend, Robert Eaglestone and Gail Vines. Dr Tristram Wyatt of the Animal Behaviour Research Group at Oxford was crucially helpful on some vital points. Deepest thanks to the Darwin Correspondence Project, to Dr John van Wyhe's Darwin Online, and to the librarians of the London Library and of the Cambridge University Library, for their invaluable help and resources.

William was this book's first and last reader, and it would never have happened without him. It is dedicated with love to Towser, Jason, Blue, Muddy, Bess, Magic, Oakley, Flash, Spuddo, Harry, Lola, Ky, Wolfie, Barney, Tallulah, Cookie, Dolly, Whistle, Beau, Cracker, Cash, Spooky and Mr Mies. And all their owners.

For more images, ideas and dogs visit www.darwinsdogs.com